INVESTIGATING LOCAL KNOWLEDGE

Investigating Local Knowledge

New Directions, New Approaches

Edited by

ALAN BICKER
University of Kent at Canterbury, UK

PAUL SILLITOE
University of Durham, UK

JOHAN POTTIER
University of London, UK

Routledge
Taylor & Francis Group

LONDON AND NEW YORK

First published 2004 by Ashgate Publishing

Reissued 2019 by Routledge
2 Park Square, Milton Park, Abingdon, Oxon, OX14 4RN
52 Vanderbilt Avenue, New York, NY 10017

Routledge is an imprint of the Taylor & Francis Group, an informa business

Publisher's Note
The publisher has gone to great lengths to ensure the quality of this reprint but points out that some imperfections in the original copies may be apparent.

Disclaimer
The publisher has made every effort to trace copyright holders and welcomes correspondence from those they have been unable to contact.

A Library of Congress record exists under LC control number:

ISBN 13: 978-0-8153-8984-2 (hbk)
ISBN 13: 978-1-138-35618-4 (pbk)
ISBN 13: 978-0-429-19938-7 (ebk)

Contents

List of Figures

List of Tables

List of Contributors

Christoph Antweiler
FB IV-Ethnologie, Universität Trier, D-54286 Trier, Germany.

G.S. Aurora
Institute for Social and Economic Change, Nagarabhavi, Bangalore – 560 072, India.

Julian Barr
Natural Resources, Environment and Social Development Work Group, Information, Training and Development (ITAD) Ltd, Lion House, Ditchling Common, Ditchling, Hassocks, West Sussex, BN6 8SG, UK.

R. Baumgartner
Nachdiplomstudium für Entwicklungsländer (NADEL), ETH Zentrum, VOB B 14, CH-8092 Zürich, Switzerland.

Jeffery W. Bentley
Agricultural Anthropologist, Casilla 2695, Cochabamba, Bolivia.

Alan Bicker
Anthropology Department, University of Kent, Eliot College, Canterbury, CT2 7NS, UK.

Eric Boa
CABI *Bioscience* UK Centre (Egham), Bakeham Lane, Egham, Surrey, TW20 9TY, UK.

Ben Campbell
Department of Social Anthropology, University of Manchester, Roscoe Building, Oxford Road, Manchester, M13 9PL, UK.

Subhadra Mitra Channa
Department of Anthropology, University of Delhi, Delhi 11000, India.

David Ellis
Anthropology Department, University of Kent, Eliot College, Canterbury, CT2 7NS, UK.

G.K. Karanth
Sociology Unit, Institute for Social and Economic Change, Nagarabhavi, Bangalore
– 560 072, India.

Johan Pottier
Anthropology Department, School of Oriental and African Studies, Thornhaugh
Street, Russell Square, London, WC1H 0XG, UK.

V. Ramaswamy
Sociology Unit, Institute for Social and Economic Change, Nagarabhavi, Bangalore
– 560 072, India.

Trudy Sable
Labrador Project, Gorsebrook Research Institute, Saint Mary's University, Halifax,
Nova Scotia B3H 3C3, Canada.

Indah Setyawati
Social Development Expert, Asian Development Bank – Indonesia Resident
Mission, Gedung BRI II, JL. Jend. Sudirman kav. 44–46, Jakharta 10210, Indonesia.

Paul Sillitoe
Anthropology Department, University of Durham, 43 Old Elvet, Durham DH1
3HN, UK.

John Stonehouse
Department of Environmental Science and Technology, Imperial College London
Royal School of Mines Building, Prince Consort Road, London, SW7 2BP, UK.

Sebastian Taylor
Polio Eradication Programme, World Health Organisation, Avenue Appia, 27
Geneva CH 1211, Switzerland.

Andrew P. Vayda
Department of Human Ecology, Cook Office Building, 55 Dudley Road, Rutgers,
The State University of New Jersey, New Brunswick, NJ 08901–8520, USA.

Percy Vilca
Agroeconomist. Jr. Nicolas de Pierola 293 RPB. Lima 6 (Dist. Carabayllo), Peru.

Bradley B. Walters
Geography Department, Mount Allison University, Sackville, N.B. E4L 1A7,
Canada.

Paige West
Department of Anthropology, Rutgers University, 131 George Street, New
Brunswick, NJ 08901, USA.

Acknowledgements

This volume was spawned by the Association of Social Anthropologists' conference on 'Indigenous Knowledge and Development' at the School of Oriental and African Studies, University of London, in 2000. We would therefore thank its contributors for their commitment to the issues raised, and Ashgate for their support throughout.

Preface

It is increasingly accepted that local knowledge (often called indigenous/traditional knowledge) has a part to play in development interventions, but its role is ambiguous. Its application in development is largely seen as assisting in the tackling of technological problems. But drawing on local knowledge equally raises political issues. What should be the terms of engagement? We know that local knowledge has weaknesses in development contexts as well as strengths, both of which result from its locally situated character. It does not necessarily comprise a comprehensive knowledge system and not all members of a community may equally share it. Consequently, even when interventions take local knowledge into account, not all may stand to benefit, and such interventions may engage in activities that are not necessarily either socially just or sustainable. What are the implications of taking part in such work? We assume that local knowledge should be the basis for building local capacity and competence, and that it should be applied as a counter-model to global science. Local knowledge needs to interface with global scientific knowledge, each drawing on the other to effect sustainable adaptation to changing natural and socio-economic environments. But engagement with development implies that external institutions, both national and international, should intervene in the activities of local people. On what grounds should we involve ourselves in such interventions? It is questions such as these to which the contributions in this volume address themselves.

Local knowledge reflects many generations of experience and problem solving by humans around the world, all increasingly affected by so-called 'globalising' forces. Development agencies are becoming aware that such knowledge, whether it be of biodiversity and ecology, natural resources management, health and disease, education and urbanisation, is far more sophisticated than development previously assumed – although anthropology has long known it. Local knowledge represents, as all the studies in this volume show, an immensely valuable resource that provides us with insights on how communities interact with their ever-changing environments, and a growing number of governments and international development agencies are recognising that local-level knowledge and organisations offer the foundation for new participatory approaches to development that are both cost-effective and sustainable, and socio-culturally sound. We find that agencies increasingly accept that the incorporation of indigenous knowledge into programmes and projects will advance development agendas, as some of the contributions here point out – indeed it is the motivation behind them, to further this trend.

This volume provides a timely overview of new directions and new approaches to investigating the role of local communities in generating knowledge based on a sophisticated understanding of their environments, devising mechanisms to conserve and sustain their lifeways, and establishing community-based organisations that serve as forums for identifying problems and dealing with them through local-level experimentation, innovation, and exchange of information with other societies. These studies show that development activities that work with and through local knowledge and organisations have several important advantages over projects that operate outside them.

Local knowledge research promotes grassroots decision-making, at the community level through indigenous structures. But it is not socially neutral. Its investigation is contentious. Any investigation of local knowledge necessarily involves consideration of intellectual property rights and other regimes concerning the protection of traditional local knowledge, methods for compensating peoples for sharing their knowledge and for protecting them against unfair exploitation. Other issues include the relation of local knowledge to the preservation of cultural and biological diversity, failures of traditional practices to maintain ecosystem health and meet human needs, methods and ethics for investigating local knowledge. Also, the relationships between traditional knowledge and Western science and their complementary use for planning and decision-making, the close involvement of local communities with development planning, the advancement of formal and non-formal education systems for the transmission of traditional knowledge, and strategies for empowering local communities and indigenous peoples to strengthen and incorporate their own belief systems into their self-determined development. Sable explores these issues in her contribution. She is concerned with how systems of education verify indigenous knowledge. She argues that the criteria for success in any international development project require nothing less than ensuring cultural continuity in the affected communities while promoting change. The notion of cultural continuity has to do with personal and cultural identities, how truths are negotiated, how knowledge is attained, and how that relates with power. What knowledge is important to people to ensure their physical, psychological, cultural, and spiritual survival, and what fundamental truths inform that choice and sustain cultural identity?

While not all of the contributors to this volume are anthropologists, their contributions have clear connections with ethnographic work. It serves to remind those involved in development that anthropology should increasingly have a role in development practice, not merely as frustrated post-project critic but as an implementing partner. It illustrates the growing need for anthropological skills and insights to further understanding of environmental, health educational, community and other issues, and so contribute in the long-term to positive change. The continuing debate about its constitution notwithstanding, development has already swallowed down the notion of indigenous knowledge, and having appropriated it, is busy adapting it to its perceived demands, modifying it with scientific perspectives, refashioning and regurgitating it to serve the needs of project and programme implementation. It has done so with little regard for the adequacy of the methods employed. Such work is not straightforward and we have a number of methodological issues to resolve to see such collaborative work become 'routine' in development contexts (Sillitoe 2002). There are a range of methodological issues to which we need to pay attention, as we have said elsewhere (Sillitoe, Bicker & Pottier 2002; Bicker, Sillitoe & Pottier 2004). We should include the following among the most pressing problems that face us in local knowledge research, to which the various contributions in this volume address themselves.

Sillitoe and Barr identify the facilitation of meaningful communication between development personnel and local people as one of the key problems with current local knowledge research. This concern is shared by Baumgartner, Karanth, Aurora and Ramaswamy in their chapter on the promotion of empowerment in rural India. They share with us the practical experience and lessons learnt from providing research feedback in the context of participatory field studies, and expose both its scope and limitations in the promotion of mutual learning. Antweiler proposes a novel model in the context of urban planning to facilitate such exchange of information. His contribution concerns the use of urban local knowledge as a way of assessing needs and improve urban life in poor countries. Deploying recent advances in cognitive anthropology together with insights from the psychology and sociology of knowledge, he has generated a model of local knowledge focussing particularly on urban knowledge that may serve as a methodological basis for the deployment of local knowledge in development.

Vayda, Walters and Setyawati argue that anthropologists have mystified indigenous or local knowledge systems and thereby exaggerated the difficulties associated with making them more widely accessible, particularly for economic development and environmental conservation. They argue that in part this reflects a preoccupation with understanding indigenous knowledge in its entirety, as a system, rather than those particular aspects of knowledge that influence behaviour relevant to any development initiative and how and why these do so.

Drawing upon work in Papua New Guinea, Ellis and West demonstrate the importance of local history in conceptions of indigenous knowledge, and in the practice of development and conservation. While emphasising the need for a broad definition of local history in relation to debates on 'local knowledge', they suggest that local history is far more diverse than biologically or economically framed representations of people, 'culture' and 'knowledge' allow. The task of sympathetically accessing concepts in local usage, and conveying something about them, is large. There is often no consensus among the 'natives', local stakeholder knowledge is rarely homogenous. People are unfamiliar with expressing all that they know in words. They may also carry knowledge, and transfer it between generations, using alien idioms. In his chapter, Campbell questions development's ability to imitate the music industry and appropriate and consume indigenous knowledge in its appetite for new techno-ethno directions. Instead he shows how a genuinely anthropological approach to knowledge-participation requires a challenging engagement with indigenous notions of identity, power and agency, which necessarily questions the terms of development participation. As he astutely notes, development's curiosity with indigenous knowledge may reflect the contemporary global consumer vogue for all things indigenous, or as he puts it, as, 'The authentic appeals of the sounds of local cultures compete with the global techno-pulse of the millennial moment'. The dynamism of local knowledge makes for difficulties in representing it too. It is subject to continual negotiation between stakeholders. And development aims to accelerate change, dramatically modifying local knowledge with scientific perspectives.

The need for interdisciplinary collaboration in local knowledge research is beautifully illustrated by the contribution of Bentley, Boa, Vilca and Stonehouse – anthropologist, tree pathologist, forester, development extensionist. They argue that local research station technicians and extension workers (*técnicos*) are a neglected source of local knowledge. Overlooked in the formal development literature, they reveal *técnicos* as experts in local conditions and indigenous rhetoric who are not only information brokers from scientists to farmers, but professionals who can explain folk concepts in scientific terms. They provide a perspective by which we can compare farmer and scientific knowledge, offering the prospect of new methods for rapid inventories of local knowledge.

We have to learn to work with contradictions, such as working with both a broad and narrow perspective, in indigenous knowledge work in development contexts, as Taylor's contribution makes clear. Starting from the premise that indigenous knowledge is, *de facto* ambiguous, he examines how local knowledge and international organisations interact, concluding that inevitably the contemporary discourse of indigenous knowledge is characterised by contradiction. He posits that locally 'indigenous knowledge' for the most part expresses itself in immediate practice. Conversely, grand international organisations experience indigeneity and knowledge as more distant, discrete concepts that they can generalise into a global-conceptual model. But development interventions, driven by a largely materialistic Western view of human advancement, often threaten the integrity of other cultural traditions. Threats of this kind have been evident since at least the start of Western colonialism, and local populations have varyingly combated them, but they have become progressively more overwhelming, until now, with the identification of the process as globalisation, many people perceive of them as accelerating out of control, indiscriminately crushing cultural differences. These issues are taken up in Channa's contribution. She focuses on issues of power in her investigation of changes in perceptions of indigenous medicine in India since independence. She describes how the upper classes are rejecting the hitherto 'superior' western scientific medical knowledge, turning instead to more indigenous systems of knowledge of disease and cure. Although discredited as inferior by the former colonial power, Channa concludes that we need to understand indigenous knowledge not only as knowledge but also as a symbol of nationhood.

Finally, the case studies presented in this volume show that in seeking to accommodate the dynamic nature of local knowledge, development interventions and indigenous knowledge research must be considered together. This edited volume makes a bold contribution towards such a strategy.

Chapter 1

Local Knowledge Theory and Methods: An Urban Model from Indonesia

Christoph Antweiler

Seek simplicity, but distrust it.
Georges Devereux (1967)

Indigenous or local knowledge is both universal and specific and defies any simple essentialism. Local knowledge is neither indigenous wisdom nor simply a form of science, but a locally situated form of knowledge and performance found in all societies. It comprises skills and acquired intelligence responding to constantly changing social and natural environments. The situated, systemic character and inherent variety of local knowledge demand multi-focussed accounts. This paper argues with a general model of local knowledge and an urban example for the importance of an explicit methodological basis of the use of local knowledge in development.

The application of local knowledge in development is less a technological than a theoretical and political problem. To make local knowledge truly relevant to development, we need more methods that allow 'thick' contextualized descriptions of such knowledge, which are also comparative and allow research on large populations. Participatory methods are useful but they generally result from tinkering and are based on a mixture of methodological aims and political ideals. To be serious and effective, methods should be based on an explicit theory of local knowledge which goes beyond simple assertions that it is complex, variable, wise and sustainable. The model of local knowledge presented is based mainly on recent cognitive anthropology, added by insights from psychology of knowledge and sociology of knowledge.

Urban knowledge is understudied in the local knowledge field and only recently used to assess needs and improve urban life in poor countries. This paper draws on an example of urban knowledge to discuss general theoretical, methodological and applied issues. The selected case is from Indonesia, a country with a developmentalist ideology, strong urban local government and a lot of old and new participation talk. I show that theoretically grounded, yet carefully adapted methods are a fruitful way for local knowledge research. In this vein, methods from cognitive anthropology are a good basis to develop participatory methods in urban areas with their high heterogeneity and ambiguity (e.g. in cultural and land tenure terms) and the importance of local government and planning. It is concluded that a systematic but localized approach is useful for the proper and humane use of local knowledge in development and for encouraging alternative forms of modernity.

1

This chapter uses an anthropological study of urban residential knowledge in the multiethnic provincial city of Makassar, South Sulawesi, to show the potentials and difficulties in empirical local knowledge research. The applied question is how can we use the knowledge and sentiments of urban dwellers in participatory planning of public and residential areas? One method seldom used so far in anthropology and development work that has considerable potential is the *repertory grid technique* originating in constructive psychology.

Participation in development, the theme of this book, requires two things: (1) a political commitment to the ideal that people should decide themselves about development aims and measures (cf. Arce and Long 1999); and (2) knowledge that is relevant to real life, real time and real space. These are necessary for participatory development, be it in poor or prosperous countries or regions. This paper mainly concerns the second knowledge requirement of participation in development. If we aim at an enablement of people in market, political and community terms (Burgess et al. 1997: 140), we need what, in the context of sustainable development in modern industrial societies, has aptly been called a 'citizen science' (Irwin 1995). The paper aims to strengthen approaches that try to use scientific knowledge, layman's knowledge and contextualized local knowledge in combination (Antweiler 1998). Beyond pure knowledge, it stresses people's ways of knowing and their experiences, sentiments and loyalties of belonging (Abram and Waldren 1998; Lovell 1998), often overlooked in local knowledge studies.

Current research into and use of local knowledge in development encourages many distortions. Local knowledge cannot be grasped by using simple dichotomies such as 'science versus wisdom', 'science versus belief', 'scientists versus folk' or 'ethnoscience versus technoscience'. These binary contrasts import hierarchies and essentialist and often orientalizing distortions. Local knowledge should not be thought of as a kind of science. On the one hand, 'Western science' comes in many varieties; on the other hand it subscribes to specific ideals and criteria (e.g. of validity and parsimony). Local knowledge may share some features with science and some with other ways of knowing. We must think of it as resulting from a universal human capability, but having specific culturally and environmentally situated content in every instance.

It is important firstly to listen to people. In addition we need direct, systematic and comparative yet culturally sensitive methods to capture the situated character and variety of local knowledge. Knowledge may be stored in people's heads and written documents, in routine practices and in material objects. Such methods are necessary to reveal common patterns among bodies of local knowledge. Furthermore, these methods should allow the study of larger populations, if local knowledge is to be relevant to development measures or counter-development.

After discussing basic conceptual ambiguities around the notions of indigenous knowledge and local knowledge (1) an example of urban knowledge is introduced (2). Based on that and the few theoretical works available a general model of local knowledge is presented comprising ten interrelated characteristics (3). Using this model several straightforward anthropologically tested methods are presented to supplement current participatory methods (4). Finally, technical and cultural adaptations of a textbook method are discussed (5).

1. Local knowledge: an ambiguous and ideologically loaded concept

Local knowledge certainly has a potential for use in development contexts (Honerla and Schröder 1995; Pasquale et al. 1998; Sillitoe 1998a), but its concrete mobilization in development is ambiguous. Theoretical, judicial-ethical and practical problems abound. Firstly, there is the unresolved epistemological status of local knowledge (Agrawal 1995; Antweiler 1998). The diverse terms and abbreviations for it reflect several – and often implicit – epistemological assumptions and diverse political agendas (table 1.1). Each of the terms has its drawbacks, but I prefer 'local knowledge', because it connotes one important dimension, situatedness in local culture and environment and avoids some connotations of other terms.

Knowledge pertaining to facts as well as skills and capabilities are local to the extent that they are acquired and applied by people with respect to local objectives, situations and problems. This knowledge is also local in that practices, i.e. problem solutions, draw on locally available raw materials and energy sources. 'Local' here is not to be understood in a strict sense as referring to a location, but rather as knowledge being culturally and ecologically situated. Local knowledge may also relate to larger regions; for instance among people that move on a routine basis (nomads, commuters, seasonal migrants), such knowledge might refer to routes or several locales.

Term	Semantic Stress Salient aspect, connotations, implicit agenda and implied opposite
• *Indigenous knowledge* (currently the dominant term)	Culturally specific self-contained knowledge, knowledge defined in relation to western (scientific) knowledge; contrast implies many dichotomies (e.g. us/them, west/rest, rationality/ magic, universal/particular, tradition/modernity)
• *Knowledge of indigenous peoples*	Knowledge of autochthonous, non-dominant or marginal, mostly non-western groups; often implies partisanship with those peoples
• *Indigenous science*	Rational, empirical knowledge; similar to western science; implies potential de-localisation
• *Local knowledge, art of locality*	Knowledge and skills related to a specific place, immediate surroundings; implies limited wider relevance
• *Endogenous knowledge; connaissance endogène locale* (fr.)	Of internal origin, rooted in local culture, (vs. exogenous respectively external knowledge)
• *Autochthonous knowledge*	Of internal origin, culturally integrated

• *Sustainable knowledge*	Environmental wisdom that provides sustainable solutions
• *Traditional knowledge*	Handed down, old, orally transmitted (implying static and homogeneity), vs. modernity, hybridity and dynamics
• *Native knowledge/expertise*	Closeness to nature, implies knowledge of a natural character
• *People's knowledge, people's science*	Shared, broadly disseminated knowledge, potential for political resistance (vs. dominating state knowledge)
• *Folk knowledge, folk science, folk competence*	Traditional, peasant, rural (in industrial societies), implies Europe's own indigenous 'other'
• *Farmers' knowledge*	Knowledge relating to the farm as an economic unit
• *Peasant knowledge*	Dominated knowledge; implies dependency
• *Partisan knowledge*	Passionate interest in particular outcome (vs. generic knowledge)
• *Little tradition*	Oral knowledge, (vs. great tradition, book knowledge)
• *Community knowledge*	Shared knowledge in self-contained social units
• *Ethnic knowledge*	Related to 'we'-group (ethnic group) and collective identity (ethnicity)
• *Cultural knowledge, cultural cognition* (in the restricted sense)	Culturally integrated and practice-oriented
• *Culturally specific knowledge*	Specificity of one group, singularity, particularity
• *Ethnoscience, ethno-science* (previously also used to denote early cognitive anthropology as a research field)	Systematic character of native models; mostly taxonomies; examples are: ethnobiology (ethnobotany, ethnozoology), ethnomedicine, ethnopharmacology, ethnoepidemiology
• *Indigenous technical knowledge*	Universality, decontextualisation and packaging possible
• *(Cultural) knowledge system*	Systematic character, generating rules (if x then y) and structures

• *(Cultural) belief system, (cultural) meaning system*	Means the same as "knowledge system", but implies a less scientific character
• *Everyday knowledge, mundane cognition, vernacular, common sense,*	Informal, applied (vs. academic, specialist, expert knowledge)
• *Know-how, savoir faire, arts de faire (fr.) practical skills,*	Praxis-oriented, action-oriented knowledge (vs. factual knowledge)
• *Performance, performance knowledge*	Action-oriented capabilities, skills for needs of the moment
• *Experiential knowledge, wisdom*	Related to (body) experience, (vs. theoretical knowledge or mere speculation)
• *Embodied knowledge*	Knowledge in the movement of body, by "acting selves"
• *Science concrete* (fr.)	Based on everyday life, on observation, (vs. *science*, fr.)
• *Situated knowledge*	Knowledge as contextual skill and 'social product'
• *mêtis* (gr.), 'a knack'	Cunning intelligence responding to constant change (vs. rule of thumb, vs. routine)
• *The science of muddling through*	Educated guesswork, trials, piecemeal incrementalism (vs. rational planning in large scale policy)
• *Experimental knowledge*	Trial-and-error, small scale trials (vs. controlled experiment)
• *Abbreviations (e.g. IK, LK, ITK, TEK, IAK, RPK)*	Modernity, systematic and scientific character, control, institutionalisation, universal character

Table 1.1   Local knowledge terms: semantic analysis

Such knowledge is most often called *indigenous knowledge*, but that sees it primarily in relation to, or in opposition to, Western scientific knowledge. This notion restricts the knowledge to specific peoples and implies uniqueness with the problems of essentialism, hierarchization, orientalism (and occidentalization of Western science). Furthermore, terms like indigenous knowledge and traditional knowledge suggest stasis, romanticism and aboriginality, whereas it comprises a pastiche of transmitted knowledge and recent invention (Sillitoe this volume). These terms invite deeply ingrained idealizations as well as negative prejudices (examples in Nygren 1999: 271–276). The term *indigenous* itself is vague and politically loaded. Today the term *indigenous peoples* is used frequently in international documents and it is impossible to use indigenous in any morally neutral or apolitical way (Ellen and Hollis 2000: 3). Many environmentalists and some social movements portray indigenous peoples as 'eco-saints'. Some see their knowledge as an antidote to some of the world's problems. In development circles it is often equated with sustainability (cf. Murdoch and Clark's 1994 'sustainable knowledge'). These idealizations are as distorting as earlier views of the 'primitive mind' or 'traditional mentality' seen as irrational and thus as a scapegoat for underdevelopment. Another danger is that terms like indigenous or local knowledge will suffer the fate of the increasingly hollow word *participation* in the development scene, with which they are currently associated.

In addition to these terminological problems revealing theoretical ambiguities there are real-world dilemmas about authorship, access and control, and of the management of knowledge. These are partly similar to those in any institution or community producing knowledge, be it a clan, a university or a software company. We depend on producing and disseminating knowledge, but must ensure that in some respects it remains the owner's property (Harrison 1995 for examples). Currently we face many difficult juridical and ethical problems of the ownership of knowledge, which are unresolved (see Greaves 1994; Posey and Dutfield 1996; Brush and Stabinsky 1996; Shiva 1997; Agrawal 1998; Strathern et al. 1998).

The application of local knowledge in development is less a technological than a theoretical and political problem. Local knowledge is often talked of and idealized by development experts as well as their critics, be it as other people's 'science' or as 'true wisdom'. Local knowledge has strengths and weaknesses in development contexts (DeWalt 1994; Antweiler 1998), both due to its culturally as well as locally situated character. The case from Indonesia presented here demonstrates the complicated nature of local knowledge, its potentials within development and the theoretical challenges.

2. Citizens as experts: urban knowledge in Indonesia

'Partisipasi' in a developmentalist state

Indonesia is a country experiencing high rates of intra-national migration and urbanization (Nas 1995; Hugo 1981, 1997; Tirtosudarmo 1997) resulting in urban problems. It also has a state ideology (*Pancasila*[1]) that is heavily oriented towards modernization. 'Development' (*pembangunan*, lit. 'awakening' or 'building up') has a very positive connotation for many Indonesians – despite their often negative experiences. There is a long tradition of community self-help (*gotong-royong*) that has been implemented in the form of unpaid cooperative labour by colonial and post-colonial governments. Since 1965 under Suharto's 'New Order' (*Orde Baru*) participation was more economic instead of political. It meant that a limited number of people participated in economic well-being, that is, an economic instead of a political or a community participation. There is a specific Indonesian tradition of 'ritualistic participation' (Schmit 1996: 195). Recently Indonesian development discourse has been influenced by participatory terminology (e.g. Quarles van Ufford 1993; Weber 1994) and versions of Western planning methods (e.g. ZOPP of GTZ).

Urban development is still heavily centralized and bureaucratic, despite recent changes in many laws and procedures. Participation is a central pillar of Indonesian urban planning and urban service programmes since 1978–9 (Karamoy and Dias 1982; Sudarmo 1997: 237), but the implementation of *partisipasi* is often heavily top-down: the people are still objects, not yet autonomous subjects of planning.

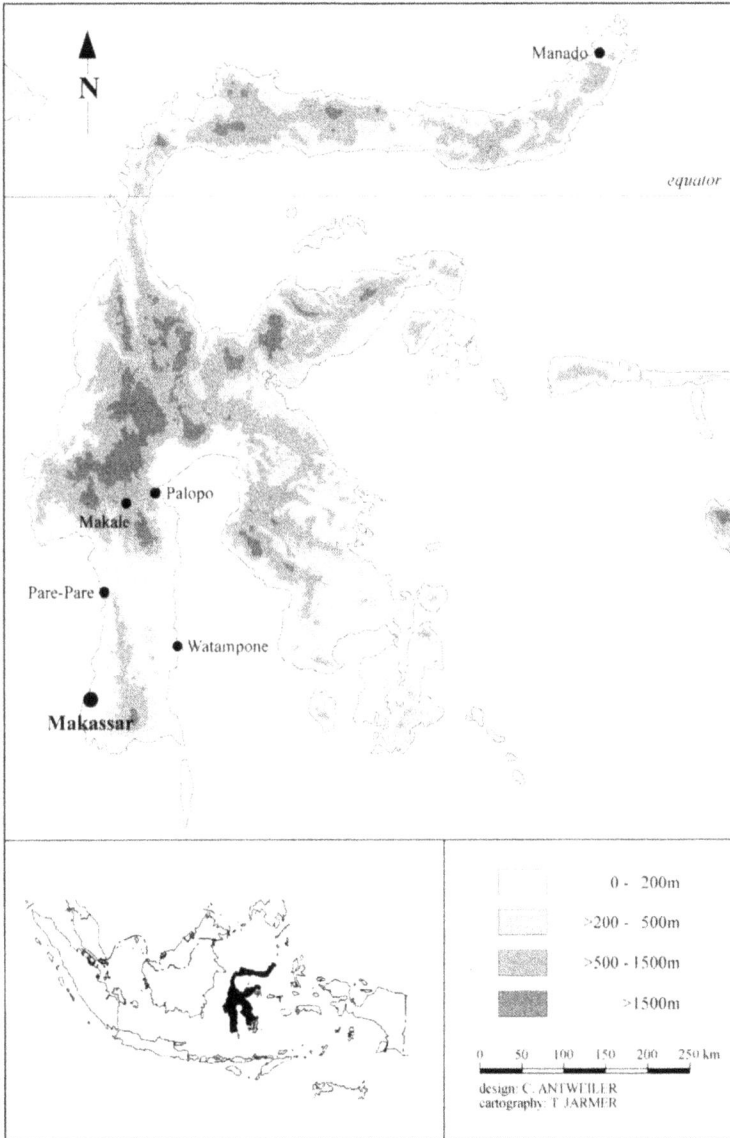

Figure 1.1   Island of Celebes (Sulawesi) in Indonesia with Makassar (Ujung Pandang), capital of *Sulawesi Selatan* province

Investigating Local Knowledge

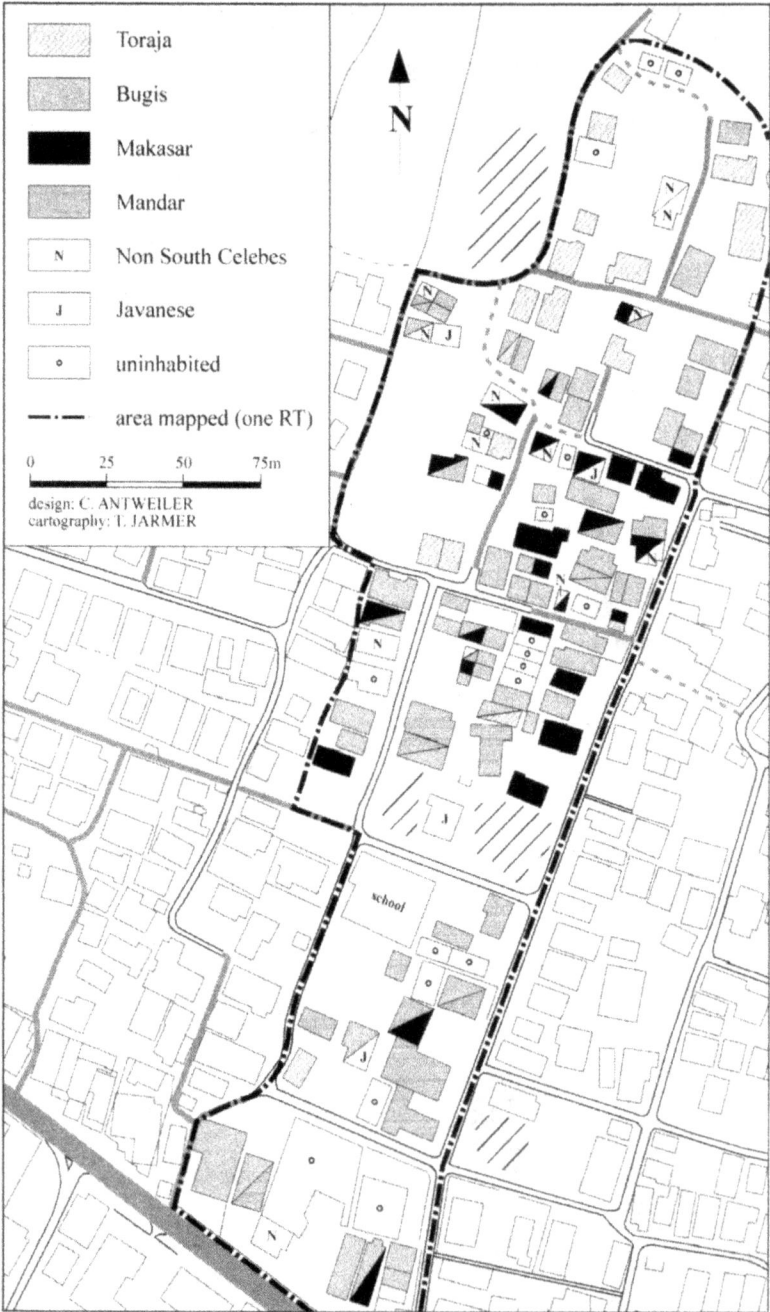

Figure 1.2   A typical ethnically mixed neighbourhood in Makassar

During my fieldwork I tried to understand local urban knowledge in relation to everyday decisions and actions. Fieldwork was conducted in Makassar (previously Ujung Pandang), a provincial city of about 1.3 million people on Celebes (*Sulawesi*) in Eastern Indonesia (figure 1.1), in an ethnically and economically mixed settlement (figure 1.2) on the outskirts called Rappocini.[2] The domain of cognition selected related to intra-urban residential mobility, which was highly relevant to informants. I related the domain to the social history of the settlement and the region, where migration has been important for a long time. Regarding methods, I conducted a detailed household survey, mapped the residences, documented residential turnover, made a residence history analysis, interviewed recently moved households and moved myself once within the neighbourhood. Several specific methods used for the study of decision making in intra-urban moves were derived from the 'natural decision making approach' in cognitive anthropology (Fjellman 1976; Gladwin 1989; Antweiler 2000: 91–126). These methods were supplemented by regular visits to building sites, empty flats and buildings, where I observed and talked with people. Specific procedures to elicit residential cognition will be discussed later in the paper.

Local mobility related concepts and strategies

People in Makassar have intricate concepts and action strategies regarding migration, residential mobility and housing (Antweiler 2000: 329–413). Figure 1.3 gives a simplified overview of the concepts and how they relate to one another. Interviewees differentiated between eight ways of earning one's livelihood, some of them linked to migration strategies. They distinguished among several house types and, in addition, identified six ways of building a house, which are associated mainly with income patterns. Furthermore, they had typologies of migration and migrants, which were linked mainly to normal male migrant biographies. Interestingly, many of these concepts are quite different from the official concepts used in maps and planning documents. The idiom is quite different from the language used in urban upgrading programmes (e.g. *Camping Improvement Programme*; *KIP*) and the programme aiming at decentralization of planning, programming and implementation (*Integrated Urban Infrastructure Development Programme*; *IUIDP*).

I will give some examples of concepts linked to residential decision making. While the people distinguish several forms of dwelling and recognize several strategies to building a house, there are only two to three in the official maps, labelled as 'permanent' (*permanen*), 'semi-permanent' (*semi permanen*) and 'temporary' (*sementara*). Whereas there are at least seven emically-differentiated ways of living in a dwelling (figure 1.3), officially there are only three: 'possessed' (*milik*), 'hired' (*sewa*) and 'staying with' (free lodging, *numpang*; e.g. In Identifikasi Kawasan Kumuh Perkotaan 1991: 6). Whereas official documents speak of illegal settlements (*illegal*) or dwellings without permission (*tanpa izin*), the people speak of 'guarding (other people's) land' (*jaga tanah*). Regarding residential mobility, the consideration of a household's or individual's 'living situation' (*situasi hidup*) was emically of central importance. This involves more than merely the economic situation and comes close to a Popperian concept of situational logic (cf. Prattis

1973). It is a concept not reflected in the official development standards of minimum wage, calorie intake etc. Exemplary local categories of *situasi hidup* are 'people seeking work' (*cari kerja*[3]), 'people seeking knowledge' (*cari ilmu*), that is people coming from the countryside who are looking for education in the city, and 'people seeking experiences' (*cari pengalaman*). This latter concept is especially interesting regarding decision making. In Makassar it is the locally accepted rationale, if people try something radically new. Additionally, people use specific notions for trial-and-error behaviour. For example, people say 'try, try often, try it again' (*coba, coba coba, coba lagi*), 'search, search again' (*cari, cari lagi*) or 'just wait and see' (*tunggu saja*). In a typical mestizo Indonesian idiom some people called these strategies *sistem eksperimental*.

Residence history interviews and decision tables[4] revealed that residential knowledge in Makassar is situated in local and regional collective memories, emotionally loaded concepts and normative biographies. It became evident that it is an amalgam of empirically based factual knowledge and forms of action-related knowing. Decision tables revealed that the knowledge is systematic and complex, but not a closed and comprehensive system shared among all people. In order to find suitable methods to study such knowledge it is not enough simply to assert that local knowledge is complex, variable and adapted. It is by no means clear what we mean by indigenous knowledge (Sillitoe 1998c: 188). In contrast to some of the other contributors to this volume I assume that an explicit analytic theory of local knowledge is required to apply it to development contexts.

3. Theory: local knowledge as situated knowledge

A cultural product beyond science versus belief

Local knowledge, despite often being called a 'knowledge system', in general does not necessarily present itself as a comprehensive system. Local knowledge informed activities are not necessarily sustainable, or socially just. Such knowledge is not only cognitive, but entails emotive and corporeal aspects. It is culturally situated and best understood as a 'social product' (Antweiler 1998). It encompasses knowledge in the strict sense of shared information and 'ways of 'knowing', 'ontology', 'framing reality', 'being acquainted with' and 'bodily knowledge', which transcend the purely cognitive (D'Andrade and Strauss 1992; Kronenfeld 1996: 14–19; Strauss and Quinn 1997; Nygren 1999: 278; Friedberg 1999: 6–10; Farnell 1999; Clammer, this volume).

> The process of acquiring insight produces awareness of one's own way of life as something worthwhile, and the mastery of skills breeds self-esteem. Thus, knowledge is created in a combination of utilitarian and symbolic practice – not in a vacuum. A plant is considered edible or characterised as just grass or weed; it is important that I know the difference, and it may turn out convenient to demonstrate that I possess the knowledge (Siverts 1991: 308).

IN SITU -
ALTERNATIVES
- add a storey
- increase house horizontally
- convert empty lower space
  into a lower storey
- install partition wall
- get free lodging
- lodging of children
  elsewhere (e.g. relatives)

FORM
- staying for longer
- 'Visiting'
- 'Being called' (e.g. by
  older relatives living in
  other places/islands)

STAY

Discontent

MOVE

**TEMPORARY
MIGRATION**

**PERMANENT
MIGRATION**

**PLACE / AREA
OF RESIDENCE**
- Rappocini
- Rest of Makassar
- Sulawesi Selatan
- East Kalimantan
- Jakarta, rest of Indonesia
- other places

**LIVING
FORMS**
- 'occupy' (illegally)
- 'guard land'
- (free) 'lodging'
- 'collective residence'
- 'rooming house'
- 'renting'
- 'own house'
- in housing estate
  - 'Dinas'
  - 'Perumnas'
  - 'private estate'

**BUILDING
STRATEGIES**
- 'building provisional'
- 'building slowly'
- 'building directly'
- 'build a stilt-house'
- 'build a stone house'
- 'build and then hire out'

**HOUSE
TYPES**
- 'traditional house'
- 'modern house'
- 'Sulawesi Selatan house'

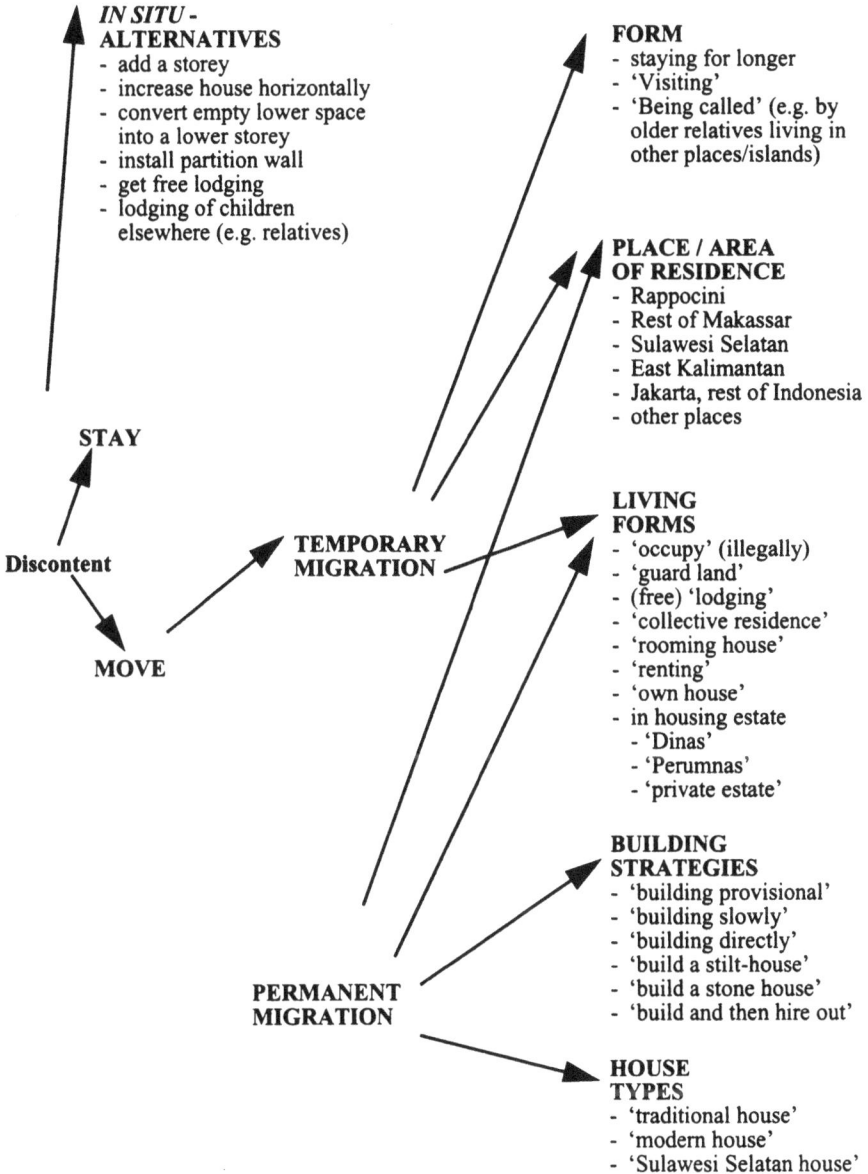

**Figure 1.3   Migration and intra-urban mobility in Makassar: a composite
model of local knowledge**

Knowledge of processes is especially relevant to development measures. Drawing on local knowledge in development should neither be restricted to the elicitation of cognitive issues or simply seen as a counter-model to Western science.

Local knowledge is frequently geared to real life practices and may only be understood with reference to the situation in which it is to be applied, called *referential transparency* (Quinn and Holland 1987). This is important in selecting research methods. The knowledge relating to practices and situations may be called *situated knowledge* (Lave 1993, 1996: 90; Nygren 1999: 277). Hobart (1993: 4, 17) speaks of *situated practices* and *knowledge as practice*. But local knowledge should not be reduced to practical, mundane, everyday or routine knowledge, as many authors do.

Patterns and varieties of knowledge, knowing and not knowing

The empirical data on urban knowledge in Makassar showed that local knowledge may on the one hand comprise fixed and structured 'knowledge' which can be articulated, or on the other hand may, by virtue of its combination with the performance of actions, involve a more fluid process of 'knowing'. Local knowledge should not be contrasted with science but may be conceived on a continuum between formal science and everyday rationality (Sillitoe 2000: 2). Furthermore, there is a distinction to be made between knowledge in the sense of *that* which is known, and knowing in the sense of *how* something is known (cf. Hobart 1993: 19; Borowsky 1994: 335–339; Barth 1995).

Local knowledge comprises not only technological and environmental knowledge, but also covers the social environment. In the final analysis, it is possible to speak in a comprehensive sense of 'knowledge of social systems' (von Cranach 1995). Local knowledge in the broadest sense also includes the social management of information, learning and teaching as well as decision making routines, given that human beings exist in a continuous flux of experiences and practices. Local knowledge can involve knowing about group peers and their interrelations, so-called 'social cognition' (Fiske and Taylor 1991; Augoustinos and Walker 1995; Pennington 2000) including the wider social environment (e.g. neighbouring communities, cf. Stokols and Altman 1987). This social cognition is especially important for development measures. Yet, it is often overlooked. Other fields worthy of mention include organizational and management knowledge, and legal knowledge, for instance pertaining to non-formal organizations (Blackburn and Andersson 1993; Marsden 1994). Local knowledge of development projects would also fall into this category, an as yet almost unstudied field.[5] Alongside this there is also more individual, less collective knowledge, for instance on how to mobilize joint interests in an action group (if there is no suitable organizational form in the local case).

Another field of political relevance is knowledge of the respective local legal system and conflict management. This knowledge is relevant when someone asks: 'How do I assert a claim?' This may require information of unwritten rules and locally appropriate ways of arguing one's case (cf. Hutchins' 1980 cognitive study of Trobriand land use conflicts and Hutchins 1996). These examples illustrate the overlaps between various fields of knowledge. Empirical studies suggest three thought levels in any knowledge: (1) recognition and naming of discreet entities; (2) ordering discontinuity and diversity; and (3) unity of ensembles of living beings, including one's own society. Distinguishing these levels may be important for eliciting parallels between scientific and local representations and establishing a dialogue between them. The chances are good for an exchange regarding the first two levels of each, but there are problems with the third (Friedberg 1999: 14).

Table 1.2 generalizes these levels of local knowledge. The examples come from both 'traditional' fields as well as modern contexts, like development, to illustrate that local knowledge is not only a traditional or rural or 'other' knowledge, but results from a universal capacity (Malinowski 1948: 196; Barth 1995; Nader 1996: 7; Scott 1998: 331; cf. Jones 1999: 559–566). This challenges the supposed contrast between 'knowledge of a different kind' (or 'non-Western knowledge' or 'non-Western wisdom') as opposed to 'Western' or 'scientific' knowledge. In real life situations, for example in development exchanges, there are usually more than two parties and two sets of knowledge.

Forms and levels of knowledge	Examples
1 Declarative knowledge	
1.1 Recognition and naming knowledge	• attribution of entities to terms, discreet entities and diversity
1.2 Factual knowledge	• traits of animals, plants, temperature, social status, prices, salaries, administrative levels
1.3 Categorical knowledge	• orderings of organisms, colours, kinship, development project types
2 Procedural knowledge	
2.1 General processes, rules	• farming calendar, religious calendar, environmental crises, household cycle, development project cycle
2.2 Specific processes ('scripts', schemas, action plans)	• everyday routines, e.g. greetings and farewells, natural resource management, ritual sequences, project request schema and non-routines
3 Complex knowledge (concepts; belief systems/ knowledge systems)	• cosmology, model of whole society, models of 'honour', of 'marriage', of 'justice', cropping systems, therapies, decision-making procedures

Table 1.2   Forms and levels of local knowledge (Antweiler 1998, modified)

Firstly, local knowledge of a declarative nature is knowledge of discreet entities relating to the natural and social environment, facts relating to neighbouring groups or, for example, details on development organizations. But local knowledge may involve categories and classifications, such as of plants, animals or relatives. The practical value of such classifications known to anthropologists for decades has only become evident in development in recent years. Ongoing debates revolve around the internal coherence and similarity to Western classifications. Disputed are also the underlying causes of these classifications: To what degree does it represent (a) pan-human perception and intellect acting on natural discontinuities; (b) utilitarian concerns; or (c) reflections of cultural relations (cf. Crick 1982: 293–298; Hunn 1982: 839–844; Berlin 1992; Ellen 1993: 3; Nazarea 1999: 4; Sillitoe 1998c)?

Secondly, local knowledge may relate to processes. This might comprise knowledge of rapid changes in the natural environment, in market prices for goods, or in experiences with development projects. Analytically we can distinguish between knowledge of general and specific processes. An example of the first would be a religious calendar; specific knowledge includes knowledge of the precise sequence of steps involved in processes (e.g. in rituals or everyday activities). There exist prototypical, idealized process models. Normal human beings, like stage performers, follow a sequence of schematically prescribed steps, ('scripts', Schank and Abelson 1977). A classic example of such an 'everyday script' is the sequence of actions performed on entering a restaurant: open door – look for unoccupied table – occupy table – look for cloakroom or hand over coat(s) – sit down – pick up menu list – order drinks etc. These 'scripts', although not fixed in writing anywhere, constitute cultural rules or sequences of steps. They are not necessarily taught, but are learned by individuals themselves through frequent performance and frequent observation. Consequently, this procedural knowledge usually remains unconscious or at least unspoken. Nevertheless, even minor errors in the sequence of actions can cause confusion among actors, or can lead to failure. If for instance I were to enter a restaurant in the United States and follow the customary European script of looking for a table myself, we would be breaking the rule of 'waiting to be seated'. Conversely, applying the American script in a European restaurant (unless it were an American Steakhouse or an exclusive) would as a rule mean waiting at the entrance to no avail.

Thirdly, local knowledge may be considerably more complex than these scripts suggest (table 1.2). It may include causal knowledge, and knowledge of complex systems, e.g. about ways of farming crops or treating environmentally induced diseases. It may relate to relationships in everyday life or information of relationships within the cosmos, or the aetiology of diseases, the creation of humankind or the origin of a community. This systemic knowledge informs action and should be considered a crucial dimension in development (cf. Warren et al. 1995). Such complex knowledge constructs may be termed 'belief systems' and knowledge systems/meaning systems (cf. van der Ploeg 1989). The word 'system' indicates that these constructs are not simply aggregates of isolated concepts, but comprise interlinked concepts and their constituent elements. Religious worldviews are an example of this. Within such systems, cosmologies and prototypical concepts play a major role. These are termed 'cultural models' or 'schemas'. Models of this kind, which may for instance centre around ideas such as 'honour and shame', 'the self-made man' or the 'typical American marriage', are usually shared by the majority. They structure knowledge processing, as they link affects and motives with accumulated experiences. They are also linked to linguistic metonymies (*pars pro toto*), analogies or visual metaphors. Although these schemes or models constitute only loose patterns of association, they are thematically organized, and are highly stable in both individual and historical terms (Quinn and Holland 1987: 24; Rogoff and Lave 1984; Strauss and Quinn 1997).

Local knowledge always has a history and is synchronically dynamic. The kernel of local knowledge consists of skills and acquired intelligence responding to *constantly changing* social and natural environments. Scott describes this masterfully with reference to the ancient Greek concept of *mêtis*, translatable as a sort of cunning intelligence, but one specifically responding to constant dynamics (Scott 1998: 311).

The diversity and dynamics of local knowledge need to be understood as part of the cultural system. There are various types of public regarding knowledge distribution. Some information is made manifest to all, whilst other information may remain concealed from the majority. Some things are known only to women or to men. Only a few specialists may possess in-depth knowledge of a particular field, for instance medical or cropping expertise. Certain individuals, or either sex, may be excluded in principle from certain fields of knowledge on cultural grounds. Even some agricultural knowledge may be categorized as secret (Pottier 1993: 30–32). The ability to know something about certain themes (*knowledgeability*, cf. Lave 1993: 13, 17) is different for the different members of a culture, and changes, as it is itself a social product. An individual's knowledge changes through time with the situation (Borowsky: 1994: 334). There may exist different or even rival items or forms of knowledge (different 'versions of the world'; Goodman 1984). As regards its content, practical relevance and centrality, knowledge is often differentially distributed between the young and the old, a process which may even lead to virtual parallel knowledge systems (Mersmann 1993). In times of rapid change, one knowledge system may be morally significant, yet without practical value. 'It is in the nature of interactions within and between "knowledge systems" to make manifest certain sorts of information and to keep other sorts hidden' (Marsden 1994: 33). Thrift (1996: 99–100) differentiates between five kinds of unknowing:

- knowledge that is unknown, because it is spatially or historically unavailable;
- knowledge not understood, that is, outside a frame of meaning;
- knowledge undiscussed and perhaps taken for granted;
- actively and consciously concealed knowledge; and
- distorted knowledge.

It is possible to draw a broad distinction between local specialist knowledge and local everyday knowledge (knowledge of 'just plain folks'; Lave 1996: 87). Studies in the sociology of knowledge and network anthropology have demonstrated that differential knowledge distribution concerns not only specialist knowledge, but also everyday knowledge. It is necessary to determine empirically who within a population at any time has what knowledge. The most intriguing issue is the attitude the people themselves have to their concepts, and the extent to which they adhere to them. What, for instance, constitutes 'factual knowledge' for the people in contrast to 'what can only be credibly claimed' for them (Sperber 1989: 98–103; Flick 1995: 56–59)?

A general model of local knowledge

How is a particular local knowledge to be defined and to what extent does it represent a 'body of knowledge'? The fields of social psychology and sociology of knowledge supply some useful definitions of knowledge. They characterize knowledge as 'stored information which refers to important structures, processes and functions of the system producing it...and which therefore generates evaluative processes' (von Cranach 1995: 25) and as 'any and every set of ideas and acts accepted by one or another social group or society of people – ideas and acts pertaining to what they accept as real for them and for others' (McCarthy 1996: 23).

An inspection of many of the recent studies of local knowledge (few of which attempt explicit generalization[6]) reveals a number of characteristic features and patterns of local knowledge. Table 1.3 outlines a general model of local knowledge, stressing both its universal capability and specific situational character. It emphasizes the practical relevance and social significance of knowledge. Local knowledge has both a practical and social dimension; it is not knowledge isolated and abstracted from everyday life. It consists not only of information, or factual knowledge as a resource, but also of capabilities and skills. The knowledge utilized for the exercise of skills remains implicit, as it is usually acquired through a process of learning by doing. With all human beings, capabilities and skills tend to be less conscious than factual and formalized knowledge. Without much comment, experienced persons demonstrate to their students how to perform actions, instead of providing detailed verbal instruction.

Key characteristic	Aspects
1 Knowledge plus skills	Combination of specific factual knowledge and practical action-oriented skills
2 Adaptation to situational dynamics and variability	Keyed to common, but never precisely identical features of a particular place; thus adapted to ambiguous, mutable, stochastic and thus in-determinant issues
3 Empirical local basis and experiential saturation	Based on local observation, low cost /-risk trial-and-error and natural experiments, proven by coping over a prolonged period in the "laboratory of life"
4 Redundancy and holism	Represented parallel in several cultural domains; embeddedness; holistic orientation through systemic relations with other aspects of culture

5 Tacit nature of knowledge	Often implicit, uncodified, intuitive, corporeal-embodied, less verbalised, less susceptible to verbal or written communication, non-disciplinary
6 Informal learning	Oral transmission, decentralised and piecemeal learning, learning by imitation, demonstration and apprenticeship more than by instruction
7 Scientific character	(Partially) systematic, methodical, parsimonious, empirical-hypothetical, comprehensive and generating causal theory
8 Optimal ignorance	Information only as detailed and accurate as it needs to address the problem; no more and no less
9 Evaluation criterion, test	Practical efficacy as the yardstick vs. e.g. theoretical consistency, parsimony, elegance etc. (but see 4)
10 Resulting actions and problem solutions	Solutions familiar and thus broadly accepted by local peoples, oriented towards 'satisficing' and optimising (vs. maximising) and the use of local/endogenous resources

Table 1.3   A general model of local knowledge: ten interrelated qualities

Local knowledge can be highly specific. Complex actions performed may depend on a procedural knowledge which is only partially conscious. This applies especially to actions involving knowledge embodied in movements (cf. Farnell 1999), rituals (Tambiah 1990) and decision making (Antweiler 2000). Procedural knowledge is less verbalized. A topical example would be asking a colleague to help with a computer problem which he or she solves after several trials but is afterwards unable to explain the steps followed. The question today is less how specific items of information find their way into people's minds, than how individuals comprise communities of knowing by participating in their knowledge. Furthermore, local knowledge is linked to some common view of the future. These may comprise 'best worlds', ontologically based theories of 'the good life' or goals of 'endogenous development'.

The combined qualities of being, both a result of universal cognitive capacities and localized/situated performance, may answer some apparent contradictions that arise if local knowledge is isolated, documented, stored, simplified, repackaged, codified, transferred and commodified (Agrawal 1995; Ellen and Hollis 2000: 17). These transformations imply de-localization, de-contextualization and de-personalization that may be possible for general know-how and skills. On the other hand, along other dimensions such as locale and environmental context, cultural embeddedness and socially situated character (sometimes featuring specified topographical information) cannot be discarded without distortion. We need systematic and comparative, yet culturally sensitive methods.

4. Methods: systematic and comparative

Participatory methods and systematic elicitation techniques

Current local knowledge data collection techniques are torn between two opposing poles. On the one hand, there are standard ethnographic methods, such as structured interviewing of cognitive anthropology (Spradley 1980; Werner and Fenton 1970; Werner and Schoepfle 1987), which are often very formal and time-consuming. On the other hand, we have the simple tools of rapid and/or participatory appraisal and learning methods (e.g. RRA, PRA, PLA, SONDEO, cf. Chambers 1991; Schönhuth and Kievelitz 1995). Both approaches have drawbacks. Cognitive anthropological methods usually only enquire into specific cultural domains in depth (e.g. the classification of animals or soils) and tend to forget people's understanding and practical use of concepts. Furthermore, often based on talking to knowledgeable specialists, intracultural cognitive variability remains largely unexplored (Sankoff 1971; Nygren 1999: 277; Sillitoe 1998c: 190). Furthermore, these methods are time-expensive. Participatory methods[7] are rapid, cheap and generally more inclusive than survey methods. But they present problems if you try to scale them up. They are theoretically ungrounded as a result of tinkering. They are laden with (often hidden) political agendas and they generally omit the larger cultural context.

This paper argues for systematic, systemic and multi-focal approaches (cf. Vayda 1983; Ellen 1996: 459). Specifically I argue that we need comparative yet culturally sensitive methods for larger samples. Methods like Rapid Rural Appraisal (RRA), Participatory Rapid Appraisal (PRA) and Participatory Learning Approach (PLA) should be complemented with procedures coming from cognitive anthropology and clinical psychology (table 1.4). The problem with cognitive methods is that while there are many works on the data processing procedures there are few on data gathering techniques in cross-cultural contexts. Anthropological experience shows that the systematic elicitation of data may be very problematic in a cultural context different from that of the researchers. The assumption of methods handbooks (e.g. Weller and Romney 1988; Werner and Schoepfle 1987) is that *systematic interviewing, systematic data collection* or *systematic elicitation techniques* may be universally applied. Cognitive anthropologists who use these methods claim to seek the emic perspective; critiques maintain that their methods are overly formal, too complicated and inapplicable to real life non-Western settings. These methods are largely based on US-American experiences with people who are accustomed to formal tests and whose cultural background is already well known. Reported experiences with such methods in the context of anthropological fieldwork in non-Western settings are varied. Some colleagues report that informants found them funny and interesting; others regarded them as childish, and instead proposed having a coffee, an *ouzo* or a *kretek* cigarette. As an illustration of the gap between bold textbook claims and fieldwork experiences compare Weller and Romney (1988) with Barnes (1991):

> ...the interviewing and data collection tasks contained in this volume are as appropriate for use in such exotic settings as the highlands of New Guinea as they are in the corporate offices on Wall Street (Weller and Romney 1988: 9).

(the informants) quickly got stuck, re-sorting the cards as each new name was added, before stopping and declaring the task to be impossible (Barnes 1991: 290).

A plea for simple cognitive methods

We may respond to these problems in three ways. Firstly, we can continue to use the methods of the established participatory approaches regardless of the problems. We should remain aware of the limitations of the methods and the fuzziness of the term 'participation'. Secondly, we can drop systematic elicitation and resort to less formal, systematic ways of collecting local knowledge data. This is the option often followed by anthropologists, collecting normal discourse, e.g. gossip. It foregoes control over data collection and results are difficult to compare.

Thirdly, we can adapt and simplify the methods of cognitive anthropology. Criticisms of cognitive anthropology mentioned above apply to older often very formal procedures (e.g. Metzger and Williams 1966). Some modern methods (cf. Weller and Romney 1988: ch.2; Antweiler 1993; Bernard 2000), such as listings, are simple and suitable if we want to know from the outset themes, criteria and problems relevant to people. Accordingly they are already used in some participatory approaches. They can elicit informants' knowledge as well as evaluations and sentiments. They allow for intracultural variation, e.g. as a consequence of age, gender, network position or social rank.

Method	**Themes, aspects (in examples)**
1. Listening, talking to people	Terminology, locally relevant cultural theme
2. Systematic/structured interviewing, formal elicitation techniques	
• free listing; question-answer-frame	domains, themes, propositions
• card/pile sorting; slip sorting	dictionary, basic cognitive structure
• triadic comparison; triad test	similarity comparisons, taxonomy
• rating, rating scale	evaluative comparisons
• ranking, rank ordering	hierarchy of values, coherence vs. diversity
• sentence frame format, frame elicitation	logical relationships, causes-effects
• combination of the above (e.g. repertory-grid method)	environmental perception, personal constructs
• graphic methods (visualisation, drawing trees, cognitive maps)	concepts; hierarchies; spatial concepts
3. Observation	
• non-participant, e.g. time allocation	practices, routines, products
• participant observation	procedural knowledge, knowing

4.	Documentation and study of documents	
	• photos, films, video	knowledge products (texts, objects)
	• mapping	knowledge distribution
	• recording of narrative texts ('orature')	nature and mode of knowledge, knowledge transfer
	• recording of natural discourse	themes, forms of discourse, implicit knowledge
5.	Combinations of the above methods	
	• natural decision-making	procedural knowledge, rules, cultural models
	• apprenticeship, teacher-pupil interaction	implicit knowledge, scripts
	• action research	knowledge acquisition, creativity, implicit knowledge
	• participatory methods (e.g. PRA, RRA, PLA)	participation in the gathering, dissemination and utilisation of knowledge

Table 1.4   Research methods for local knowledge

5. Urban knowledge and sentiment: localizing methods

Residential knowledge and repertory grids

During my research into intra-urban residential mobility specific questions arose concerning knowledge and the sentiments of residential relocation and the subjective relevance of the built environment (*residential cognition*; cf. Tognoli 1987; Saegert and Winkel 1990): What do people know of the area (*knowledge*)? How do they know it; what are the ways in which they represent streets cognitively (*knowing*)? What are their evaluations and the meanings of the built environment? Which are the evaluations of neighbourhoods or places as potential residential locations?

Environmental psychologists and geographers interested in perception often use techniques of *cognitive* or *mental mapping* to understand such issues. Months of living with people in Makassar and exchanges with a research assistant showed that they are not accustomed to using maps. Any results would have been quite artificial. I turned to the *Repertory Grid Method* (also called *Repgrid-technique*) that elicits cognitive and emotive data via language to reveal so-called 'constructs' (Scheer 1993: 25–36). The method was developed by George Kelly (1955; cf. Fansella 1995) and is classically used in clinical psychology to reveal subjective theories (Catina and Schmitt 1993). Guides to the method include Scheer and Catina (1993) and Fromm (1995). It is also used by environmental psychologists and urban geographers interested in neighbourhood evaluation and residential choice (e.g. Aitken 1984, 1987, 1990; Anderson 1990; Preston and Taylor 1981; Tanner and

Foppa 1996). The Repertory Grid Method is hardly known in anthropology and seldom used in the field of development (for exceptions see Barker 1980: 300; Richards 1980: 187; Seur 1992: 124–127).

The simple basic assumption is that humans order their world cognitively by using dual polarities. Referring to everyday decisions the assumption is that individuals 'construe' several aspects within the diversity of their experiences according to similarities and dissimilarities (Catina and Schmitt 1993). Every person uses many polarities and these differ intra and interculturally, which is different to the principal dualism assumed by structuralists. The main assumptions of the Repertory Grid Method are:

- Humans do not merely react to events but are in charge of their actions.
- People think in dichotomies of diverse content.
- Personal constructs consist of psychic polarities. Each pole is only relevant with its counterpart, but they need not be logical poles or contradictions; they need not to be rational or precise.
- Personal repertoires are rooted in the biographic framework of earlier experiences and anticipation of future.
- A person has a repertoire of several such constructs.
- Individuals select between several of their constructs according to the situation given.
- Individuals can rebuild their constructs. Alternative views are present anytime, which is important in everyday decision making.

Procedure and advantages of the Repertory Grid Method

The method consists of two steps: (1) an elicitation of constructs via a comparison of two or three items (dyads, triads); and (2) an evaluation of (other) items according to the constructs elicited in a scaling procedure.

Step 1 is the 'construct question'. Two or three items called 'elements' are compared by the person interviewed. These elements might be words or short sentences on cards, photographs or concrete objects, such as a plant or an animal specimen. The interviewer asks the subject to discriminate on the basis of similarity (*dyad comparison, triad comparison*), without any other inputs. This reveals certain stated characteristics, for example 'these two are similar because they are both clean' or 'this one is different because it is dangerous'. Thus an 'initial construct pole' emerges. Asking for the reverse of the stated trait – if not obvious – reveals pairs of contrasts (e.g. 'clean/dirty', 'dangerous/secure'). These pairs of contrasts are called 'polarities' or 'personal constructs'. It is possible to elicit several constructs per dyad or triad.

Step 2 uses the constructs elicited as poles on a scale for ordering other items presented to the interview partner. New items are evaluated by ranking or rating them on a scale established by the interviewee, not the ethnographer. Only the items for comparison come from the interviewer. This differentiates this method from *semantic differential* or *polarity profile* procedures, where given items are rated or ranked. A matrix can be formed by arranging the constructs (from step 1) horizontally and the rating values (from step 2) vertically. It represents the person's cognitive and emotive repertoire regarding a specific theme or domain. The semantic space consisting of elements and constructs is called the person's *repertory grid*.

Here I present the method in the form of a recipe adapted to the Makassar fieldwork setting (figure 1.4). Subsequently I explain the adaptation to the local setting – an important step often not mentioned in cognitive anthropology methods texts and RRA, PRA, PLA handbooks.

Step One: Elicitation of constructs via triad comparison:

1. Present three elements (houses, lanes, neighbourhoods) A, B, C as photos.
2. Ask: 'Which of these two are similar or which one is specific in any respect?'
3. Ask: 'Why this one/these ones?' Note response as one pole on a 10–point-scale.
4. Ask: 'What would be the reverse of that?' Note reply as other polarity.
5. Ask: 'Which one would you prefer of these two qualities?' Note preference with a symbol (+).
6. Repeat several triads with the same procedure to obtain several constructs.

Step Two: Assessment of named urban areas with the constructs (elicited in step 1):

1. Ranking of eight plastic slips with the names of residential areas or streets within the poles of the first polarity profile.
2. Repeat this ranking of the eight slips within the other constructs elicited in step one.

The method has qualitative and quantitative aspects and several advantages compared either to survey-like methods or open interviews. In the media used there are similarities to participatory methods (e.g. PRA), except that the method is more rigorous but not less participatory. (1) It elicits systematic emic data from a sample (same stimuli for everyone) and is thus both qualitative and quantitative. (2) We gain emic cognitive results of individuals (not generalized ones) and can document intracultural variation. (3) The researcher gives only stimuli instead of a prearranged polarity profile, thus allowing for local perspectives (e.g. local classifications or sentiments). (4) Photos are used as stimuli instead of words, thus allowing for more control of stimuli. (5) The items are presented parsimoniously instead of an elaborate procedure and are thus applicable to real life. (6) Notation is easy and simple and thus transparent for the interview partners, which is also an ethical issue. (7) The interview procedure is short and thus open for further dialogue on emically related topics. The method could be modified in many ways. We could, for example, ask people to sort or rank items according to desired future states of their residential environment (Aitken 1990: 253). The method elicits subjective theories, but being comparative oriented it may be used to understand intersubjective emic theories.

Local adaptation: culturally relevant themes and suitable media

If participation is our aim, textbook methods demand considerable modification. The challenge is to simplify and adapt the methodology without localizing it so much as to prevent comparison and generalization. The localization aspect is often forgotten in applying PRA methods, because it needs time and some ethnographic grounding as I will now show. I resorted to this method after having been in the field for five months and tried different approaches, for example cognitive maps. I interviewed a random sample of house owners (30 per cent; n=21) from the neighbourhood. My respondents determined the interview locations. Often we talked in their small guest rooms (*ruang tamu*), on the terrace, or in front of the dwelling at the edge of the small lane. Interviews lasted between 1 and $1^1/_2$ hours depending on age and education of the interviewees, and situational factors such as their mood and presence of neighbours.

introductory triad (3 persons A,B,C) to sensibilise interviewee for an evaluative similarity comparison

notation form for constructs and ranking (steps 1 and 2)

kretek-clove cigarettes

plastic slips (# 1 to # 8) for ranking names of streets or neighbourhoods and the birth-place of interviewee

triads (1,2,3 / 4,5,6 ... 19,20,21) of built environment for generating constructs (step 1)

Figure 1.4   Set of interview material for *repertory grids*

First experiences confirmed that triads motivate people to think and evaluate far more than dyads, which are often too obviously similar or different. For any triad I selected photographs that showed relatively similar situations to make the comparison interesting, but tried to maximize the variance of living situations through the triads. One triad showed three poor houses while in another the interviewee saw three well-organized middle class neighbourhoods. I used photographs (figure 1.4) for step one, the elicitation of the constructs, because I found during the first few months of my stay that people like to talk about photographs; they keep albums of family photographs and almost every household displays photos. It is relatively easy to present the same stimuli to every person interviewed using photos. Furthermore, photographs are suitable for exploring urban environmental knowledge because they reveal details of living spaces that people can compare.

The colour photographs showed typical residential areas from Makassar. The selection of photographs was important. I did not use photos of the area where the interviewees lived, because in step one I wanted to elicit general value orientations based on observable traits and not the evaluation of a known area (Nasar 1998). Likewise I did not use photos of well-known places, streets, buildings or advertising hoardings. I selected from hundreds of photos that I had taken during the first five months of fieldwork, to document the city, to cover the diversity of living situations and residential areas. Pragmatic as well as ethical considerations prompted me to show only outdoor situations. Firstly, indoor photos too obviously reflect residents' social status and thus would not be very productive. Secondly, I could only take indoor photos where I knew the people well. Using indoor photos of households personally known to the interviewees would have been dubious in a society where living conditions are a common theme of everyday gossip. Further practical considerations regarding photographs included using the locally common 10cm x 15 cm print format and numbering the photos for easy identification. This also facilitated notation and was transparent to interview partners, a significant consideration for participatory citizen science methodology.

The comparative evaluation of urban areas within the constructs (step 2) also needed some preparation based on my fieldwork experiences. For the stimuli I presented eight urban areas using their names (not photographs). The neighbourhood names were written clearly on plastic slips. They were easy to handle if there was only dim light within a house or in the evening, and they didn't get dirty if the interview had to be conducted outside in front of a house or during rain in the monsoon season. I selected six areas, an interviewee's present neighbourhood, and his or her birthplace. The reasoning behind the inclusion of the last two areas was to connect the ranking with the person's biographical experience. During my fieldwork about residential decision making it had become apparent that previous country–city migration experiences are a key factor in selecting areas for intra-urban residential moves. Awareness was to be guided towards general evaluations, images and prejudices, not specific traits. Ranking was used instead of rating because it is less laborious and implies a comparison of elements.

Systematic yet culturally adequate interviewing: experiences

As might be expected, people from an ethnically and economically mixed neighbourhood reacted differently to this method. Most of them were interested or amused, but to some it was strange. Having been in the field for five months, I had visited all interviewees several times before the repertory grid interview. I had had informal discussions, and had conducted a household census and a residential history interview with all of them. So they knew me and knew my overall research topic was intra-urban residential mobility.

When introducing this specific interview I stressed that it was not seeking 'correct' answers, that I was interested in their personal perceptions and evaluations. This was important as my trials with this formal procedure revealed that some people associated it with intelligence tests. In a city with many schools, universities and offices many people have had experiences with such tests. After some conversation and a few obligatory *kretek* clove-cigarettes I showed the people three outdoor photographs: one of myself, one of my brother and one of my mother. I used my family because people are very keen on seeing photos of Western people and of families. I then asked the triad question and answered it by grouping according to sex: my brother and myself versus my mother. Then I pointed out that we could also group in the same way but for other reasons. My mother was improperly (*kurang cocok*) dressed for women, whereas my brother and I were properly dressed. The aim was to demonstrate the method to the interviewee using an everyday topic. And I used gender roles to make them aware of an evaluative comparison in the interview. I deliberately used a very simple notation form. I used big letters because of my poor handwriting and dim light. The notation was facilitated by the numbering of the photographs (step 1) and the plastic slips (step 2).

The repertory grid method presented here is simple regarding theoretical assumptions and procedure followed, but using it requires reasoning and some preparation. Time is needed, as most of this preparatory work has to be done in the field. The conclusions drawn from the repertory grid interview in my specific case are as follows:

- Even 'simple' cognitive textbook methods are complicated and time consuming.
- A problem of cognitive textbook methods is that they often require many interviews with the same person.
- A short interview is the better one in most instances.
- Increase transparency by using simple notation allowing for visual sharing.
- Restrict the number of personal constructs elicited.
- Ensure that items presented are culturally appropriate.
- Photographs are often suitable as people are interested in them.
- Plan time to prepare suitable photos and pre-test them.
- Ask what the locally relevant themes of discourse are?
- Prepare the interview partners with another locally relevant topic using the same method.

everyday concepts		official idiom (e.g. in maps and planning documents)	
pole 1	pole 2		
(being) alone (*sendiri*)	populated (*bermasyarakat*)		
like village (*sama kampung*)	urban (*kota*)		
still like kampung (*masih kampung*)	already urban (*sudah kota*)		
plain, simple (*sederhana*)	luxurious (*mewah*)		
orderly (*rapi*)	not (yet) orderly (*belum rapi*)		
dense (population) (*padat*)	distantly spaced (*renggang*)		
dense (population) (*padat*)	good ventilation (*udara bagus*)		
dense (population) (*padat*)	still empty (*masih kosong*)		
calm (*tenang, sunyi*)	ado, duss, gossip (*cencong*)		
calm (*sunyi*)	full of life (*ramai*)		
secure (*aman*)	disturbed, unsafe (*rawan*)		
secure (*aman*), clear (*tenang*)	insecure (*kacau*)	same as everyday concept	
orderly planned (*teratur*)	not orderly (*tidak teratur*)	same as everyday concept	
the rich (*yang kaya*)	ordinary people (*orang biasa*)	rich (*orang kaya*)	poor (*miskin*)
modern (*maju*)	not (yet) modern (*belum maju*)	modern (*maju*)	traditional (*tradisional*)
dirty *(kotor)*	clean (*berskh*)	dirty area (*kawasan kumuh*)	
		economically strong (*economi kuat*)	economically weak (*economi lemah*)
		upper people (*masyarakat tinggi*)	lower people (*masyarakat rendah*)

**Table 1.5   Concepts about residential areas: emic concepts vs. official idiom
(selected concepts, empty fields: no complement observed)**

Results: the citizen's voice and the official idiom of development

The repertory grid method yields various results, ranging from simple qualitative data about cases to detailed quantitative sample data. The results may be employed in several ways:

- simple graphic representation to allow visual sharing, which is a good basis for direct further discussion and reveals proposals for as yet unrecognized evaluation criteria;
- analysis through simple sorting by hand (cf. Raethel 1993: 47–49);
- using software analysis, e.g. Anthropac (Borgatti 1989); and
- use of specific quantitative and graphic data processing (cf. Raethel 1993: 53–67) for which there is software (see Willutzki and Raethel 1993).

The results showed striking differences between the idiom and perceptions of local residents and the language and concepts employed in official urban planning brochures (table 1.5). Formal but simple cognitive methods have a potential in development as demonstrated using local urban knowledge to achieve more humanized and effective urban planning. Regarding methods, the universality of cognitive approaches can be maintained provided that either informants are accustomed to formal questioning, or textbook versions are adapted to the local cultural setting. The latter requires a certain ethnographic grounding in local culture, which is not normally available in development projects.

Conclusion: local knowledge for community enablement?

Local knowledge research in the realm of development is problem oriented, but even such research needs well-defined methodologies, which require a clear theoretical understanding of the phenomenon of local knowledge. The kernel of local knowledge should be seen in a form of knowledge and performance found in all societies, comprising skills and acquired intelligence, which are culturally *situated* and responding to *constantly changing* social and natural environments. Cognitive anthropological methods are applicable provided that textbook versions are considerably simplified and adapted to the local cultural setting. Participatory methods often used in development currently are useful but are frequently too quick and mostly lack theoretical grounding. The 'Repertory Grid Method' is presented here as a method which allows the elicitation of information regarding local cognition and emotion in a systematic yet sensitive way.

In urban situations, besides knowledge about dwellings, space and mobility options, knowledge of prices, unwritten rules of the public sphere and of bureaucracy are particularly important. Such knowledge could be used both to counter dominant official regulations and to enrich expert knowledge. If local knowledge research were less idealistic, hurried, more systematic, multi-focal and context-sensitive, this would make development and specifically urban planning more effective and participatory. The aim would be towards community enablement.

Acknowledgements

I thank participants at the ASA 2000 conference 'Participating in Development' for suggestions during the discussion of an earlier version of this paper. Further thanks go to Michael Schönhuth (University of Trier), Cornelia Sperling and especially to Alan Bicker and Paul Sillitoe, whose many critical remarks and suggestions improved the paper considerably.

Notes

1   All non-English terms are in *Bahasa Indonesia*, the national language, in the idiom colloquially used in Makassar.
2   Together with my wife and our son, then seven months old, I lived there for one year in 1991/92. During the first six months we lived in a single room of a small house inhabited by an Indonesian family with four children. The couple was ethnically mixed (her a Buginese, him a Mandarese). They normally spoke *Bahasa Indonesia*, as usual in urban households in Indonesia. The husband was a low-level government clerk (*pegawai negeri*) and his wife worked in the house. During the second half of the year we lived in a nearby neighbourhood in the same settlement with a better-off childless couple, again ethnically mixed, her a Makassarese and him a Minangkabau from west Sumatra.
3   I use the local vernacular of *Bahasa Indonesia* throughout. The correct term here, e.g., would be *pencari kerja*.
4   A 'theoretical sample' (cf. Strauss 1987) of 37 people were interviewed about their residence history. Additionally, decision tables were developed with 25 interviewees.
5   Many underpaid academics in Indonesia for example have an intricate knowledge about the ideals and procedures of northern NGOs. They use this to acquire money from them or to establish their own NGOs (*LSM, Lembaga Swadaya Masyarakat*), often as an income-generating device.
6   The following works from different disciplinary backgrounds have informed my understanding of the general characteristics of local knowledge: Malinowski 1948; Lindblom 1959; Crick 1982; Tambiah 1990; Lave 1993; DeWalt 1994; Barth 1995; Harrison 1995; Lambek 1993; Lave 1996; Nader 1996; Strauss and Quinn 1997; Antweiler 1998; Sillitoe 1998a, 1998b; Scott 1998; Friedberg 1999; Nazarea 1999; Nygren 1999; Ellen and Hollis 2000; Sillitoe this volume; Clammer this volume and Sinclair et al. 2000.
7   Participatory methods are widely used in development projects but due to their rural tradition are still comparatively seldom used in the urban sector (e.g. PUA). As an indication, there are only 136 urban items listed among over 2000 items in the Participation Reading Room of the Institute of Development (IDS) in Sussex (http:// nt1.ids.ac.uk, downloaded 22.3.2001). See the RRA Notes, No. 21 'Special Issue on Participatory Tools and Methods in Urban Areas' (Nov. 1996) for short examples and Cresswell (1996) as an instructive detailed example.

References

Abram, S. and Waldren, J. (eds.) 1998. *Anthropological perspectives on development: Knowledge and sentiment in conflict.* London, New York: Routledge (European Association of Social Anthropologists).

Agrawal, A. 1995. Dismantling the divide between indigenous and scientific knowledge. *Development and Change* 26: 413–439.

Agrawal, A. 1998. Geistiges Eigentum und indigenes Wissen: Weder Gans noch goldene Eier. In Flitner, M., C. Görg & V. Heins (eds.): *Konfliktfeld Natur. Biologische Ressourcen und globale Politik.* 193–214. Opladen: Leske und Budrich.

Aitken, S.C. 1984. Normative views and ordering the urban milieu. *East Lakes Geographer* 14: 1–16.

Aitken, S.C. 1987. Households moving within the rental sector: mental schemata and search spaces. *Environment and Planning A* 19: 369–383.

Aitken, S.C. 1990. Local evaluations of neighbourhood change. *Annals of the American Association of Geographers* 80(2): 247–267.

Anderson, T.J. 1990. Personal construct theory, residential decision-making and the behavioural environment. In F.W. Boal & D.N. Livingstone (eds.) *The behavioural environment. Essays in reflection, application and re-evaluation.* London & New York: Routledge. 133–162.

Antweiler, C. 1993. Universelle Erhebungsmethoden und lokale Kognition am Beispiel urbaner Umweltkognition in Süd-Sulawesi/Indonesien. *Zeitschrift für Ethnologie* 118(2): 251–287.

Antweiler, C. 1998. Local knowledge and local knowing: an anthropological analysis of contested 'cultural products' in the context of development. *Anthropos* 93(4–6): 469–494.

Antweiler, C. 2000. *Urbane Rationalität: eine stadtethnologische Studie zu Ujung Pandang (Makassar), Indonesien.* Berlin: Dietrich Reimer Verlag (Kölner Ethnologische Mitteilungen, 12).

Arce, A. & Long, N. (eds.) 2000. *Anthropology, development and modernities: exploring discourses, counter-tendencies and violence.* London & New York: Routledge.

Augoustinos, M. & Walker, I. 1995. *Social cognition: an integrated introduction.* London etc.: Sage Publications.

Barker, D. 1980. Appropriate technology: an example using a traditional African board game to measure farmer's attitudes and environmental images. In *Indigenous knowledge systems and development.* (eds.) D.D. Brokhensha, D.M. Warren & O. Werner. Washington D.C.: University Press of America (The International Library of Development and Indigenous Knowledge). 297–302.

Barnes, R.H. 1991. Review of Röttger-Rössler 1989 *Zeitschrift für Ethnologie* 115:289–291.

Barth, F. 1995. Other knowledge and other ways of knowing. *Journal of Anthropological Research* 50: 65–68.

Berlin, B. 1992. *Ethnobiological classification: Principles of categorization of plants and animals in traditional societies.* Princeton: Princeton University Press.

Bernard, H.R. 2000. *Social research methods: Qualitative and quantitative approaches.* Newbury Park etc.: Sage Publications.

Blackburn, T.C. & K. Andersson. (eds.) 1993. *Before the wilderness: Environmental management by native Californians.* Menlo Park: Cal.: Ballena Press.

Borgatti, S.P. 1989. *Provisional documentation. ANTHROPAC 2.6.* Manuscript. no place given.; n.y.

Borowsky, R. 1994. On the knowledge and knowing of cultural activities. In *Assessing Cultural Anthropology.* (ed.) R. Borowsky. New York: McGraw-Hill. 331–347.

Brush, S. B. & Strabinsky, D. (eds) 1996 *Valuing local knowledge: indigenous people and intellectual Property Rights.* Washington D.C.: Island Press.

Burgess, R., Carmona, M. & Kolstee, T. 1997. Contemporary policies for enablement and participation: A critical review. In. *The challenge of sustainable cities: neoliberalism and urban strategies in developing countries.* (eds.) M. Burgess, Carmona, M. & T. Kolstee. London: Zed Books. 139–162.

Catina, A. & G.M. Schmitt. 1993. Die Theorie der Persönlichen Konstrukte. In *Einführung in die Repertory Grid-Technik. Band 1: Grundlagen und* Methoden. (eds.) J.W. Scheer & A. Catina. Bern: Verlag Hans Huber. 11–23.

Chambers, R. 1991. Shortcut and participatory methods for gaining social information for projects. In *Putting people first: Sociological variables in rural development.* (ed.) M.M. Cernea. Oxford: Oxford University Press (for the World Bank). 515–637.

Cresswell, T. 1996. Participatory approaches in the UK urban health sector: Keeping faith with perceived needs. *Development in Practice* 6(1): 16–24.

Crick, M.R. 1982. Anthropology of knowledge. *Annual Review of Anthropology* 11, 287–313.

D'Andrade, R.G. & C. Strauss. (eds.) 1992. *Human motives and cultural models.* Cambridge: Cambridge University Press.

Devereux, G. 1967. *From anxiety to method in the behavioral sciences.* Den Haag: Editions Mouton & Company/Ecole Pratique des Hautes Etudes.

DeWalt, B.R. 1994. Using indigenous knowledge to improve agriculture and natural resource management. *Human Organization* 53(2): 123–131.

Ellen, R.F. 1993. *Nuaulu ethnozoology: A systematic inventory.* Canterbury: University of Kent at Canterbury.

Ellen, R.F. 1996. Putting plants in their place: Anthropological approaches to understanding the ethnobotanical knowledge of rainforest populations. In *Tropical Rainforest Research-Current Issues.* (eds.) D. S. Edwards, et al. Dordrecht: Kluwer Academic Publishers.

Ellen, R.F. & Hollis, H. 2000. Introduction. In *Indigenous environmental knowledge and its transformations: Critical anthropological perspectives.* (eds.) R.F. Ellen, P.Parkes, & A. Bicker. Abington: Harwood Academic Publishers. 1–25.

Fansella, F. 1995. *George Kelly.* London, New York: Sage.

Farnell, B. 1999. Moving bodies, acting selves. *Annual Review of Anthropology.* 28: 341–373.

Fiske, S.T. & S.E. Taylor. 1991. *Social cognition.* London: Addison-Wesley.

Fjellman, S. 1976. Natural and unnatural decision-making: A critique of decision theory. *Ethos* 4: 73–94.

Flick, U. 1995. *Psychologie des Sozialen. Repräsentationen in Wissen und Sprache.* Reinbek bei Hamburg: Rowohlt Taschenbuch Verlag.

Friedberg, C. 1999. Diversity, order, unity: Different levels in folk knowledge about the living. *Social Anthropology* 7(1): 1–16.

Fromm, M. 1995. *Repertory Grid Methodik. Ein Lehrbuch.* Weinheim: Deutscher Studien Verlag.

Gladwin, C.H. 1989. *Ethnographic decision tree modeling.* Newbury Park, Ca. & London: Sage Publications.

Goodman, N. 1984. *Weisen der Welterzeugung.* Frankfurt: Suhrkamp.

Greaves, T. (ed.) 1994. *Intellectual property rights for indigenous peoples: A source book.* Oklahoma City: Society for Applied Anthropology.

Harrison, S. 1995. Anthropological perspectives on the management of knowledge. *Anthropology Today* 11(5): 10–14.

Hobart, M. 1993. Introduction: The growth of ignorance. In *An anthropological critique of development. The growth of ignorance.* (ed.) M. Hobart London, New York: Routledge. 1–30.

Honerla, S. & P. Schröder. (eds.) 1995. *Lokales Wissen und Entwicklung.* Saarbrücken: Verlag für Entwicklungspolitik (*Entwicklungsethnologie*; special issue).

Hugo, G. J. 1981. Village-community ties, village norms, and ethnic and social networks: A review of evidence from the Third World. In *Migration decision making. Multidisciplinary approaches to microlevel Studies in developed and developing countries.* (eds.) G.F. De Jong & R.F. Gardner. New York etc.: Pergamon Press. 186–221.

Hugo, G. J. 1997. Population change and development in Indonesia. In *Asia-Pacific. New geographies of the Pacific rim.* (eds.) R.F. Watters, T.G. McGee & G. Sullivan London: Hurst & Company. 223–249.

Hunn, E. 1982. The utilitarian factor in folk biological classification. *American Anthropologist* 84: 930–847.

Hutchins, E. 1980. *Culture and inference: A Trobriand case study.* Cambridge, Mass. & London: Harvard University Press.

Hutchins, E. 1996. *Cognition in the wild.* Cambridge, Mass.: The MIT Press.

Identifikasi Kawasan Kumuh Perkotaan 1991. Ujung Pandang: Traksi Perdana Konsultan Teknik.

Irwin, A. 1995. *Citizen science: a study of people, expertise and sustainable development.* London: Routledge.

Jones, D. 1999. Evolutionary psychology. *Annual Reviews of Anthropology* 28: 553–575.

Karamoy, A. & G. Dias. (eds.) 1982 *Participatory urban services in Indonesia: people participation and the impact of government social services programmes on the kampung communities: a case study in Jakarta and Ujung Pandang.* Jakarta: Lembaga Penelitian, Pendidikan dan Penerangan Ekonomi dan Sosial (LPIIIES).

Kelly, G.A. 1955. *The psychology of personal constructs* (2 vols.). New York: W. W. Norton & Company.

Kronenfeld, D.B. 1996. *Plastic glasses and church fathers: Semantic extension from the ethnoscience tradition.* Oxford: Oxford University Press.

Lambek, M. 1993. *Knowledge and practice in Mayotte: Local discourses on Islam, sorcery and spirit possession.* Toronto: University of Toronto Press.

Lave, J, M. Murtaugh & O. de la Rocha. 1984. The dialectic of arithmetic in grocery shopping. In *Everyday cognition: Its development in social context.* (eds.) B. Rogoff & J. Lave. Cambridge, Mass.: Harvard University Press. 67–94.

Lave, J. 1993. The practice of learning. In: *Understanding practice: perspectives on activity and context.* (eds.) J. Lave & S. Chaiklin. Cambridge: Cambridge University Press. 3–32.

Lave, J. 1996. The savagery of the domestic mind. In *Naked science: Anthropological inquiry into boundaries, power, and knowledge.* (ed.) L. Nader New York & London: Routledge. 87–100.

Lindblom, C.E. 1959. The science of 'muddling through'. *Public Administration Review* XIX (Winter) 79–88.

Lovell, N. 1998. Introduction. In *Locality and belonging.* (ed.) N. Lovell. London & New York: Routledge (European Association of Social Anthropologists). 1–24.

McCarthy, D. 1996. *Knowledge as culture: The new sociology of knowledge.* London & New York: Routledge.

Malinowski, B. 1948. (1925) *Magic, science and religion and other essays.* Garden City, N.Y.: Doubleday Anchor.

Marsden, D. 1994. Indigenous management and the management of indigenous knowledge. In *Anthropology of Organizations.* (ed.) S. Wright London, New York: Routledge. 41–55.

Mersmann, C. 1993. *Umweltwissen und Landnutzung in einem afrikanischen Dorf: zur Frage des bäuerlichen Engagements in der Gestaltung der Kulturlandschaft der Usambara-Berge Tansanias.* Hamburg: Deutsches Übersee-Institut.

Metzger, D.G. & G.E. Williams. 1966. Some procedures and results in the study of native categories: Tzeltal 'firewood'. *American Anthropologist* 68: 389–407.

Murdoch, J. & J. Clark. 1994. Sustainable knowledge. *Geoforum* 25(2): 115–132.

Nas, P.J.M. (ed.) 1995. *Issues in urban development: Case studies from Indonesia.* Leiden: Research School CNWS.

Nader L. (ed.) 1996. *Naked science. Anthropological inquiry into boundaries, power, and knowledge.* New York & London: Routledge

Nasar, J.L. 1998. *The Evaluative image of the city.* Thousand Oaks etc.: Sage Publications.

Nazarea, V. D. 1999. Introduction: A view from a point: ethnoecology as situated knowledge. In *Ethnoecology. Situated knowledge / Located lives*. (ed.) V.D. Nazarea. Tuscon: University of Arizona Press. 3–20.

Nygren, A. 1999. Local knowledge in the environment-development discourse: From dichotomies to situated knowledges. *Critique of Anthropology* 19(3): 267–288.

Pasquale, S., Schröder, P. & Schulze, U. (eds.) 1998. *Lokales Wissen für nachhaltige Entwicklung: ein Praxisführer*. Saarbrücken: Verlag für Entwicklungspolitik.

Pennington, D.C. 2000. Social cognition. London: Routledge & Philadelphia: Taylor & Francis.

Preston, V. A. & S. M. Taylor. 1981. Personal construct theory and residential choice. *Annals of the Association of American Geographers*. 21: 437–461.

Posey, D.A., & G. Dutfield. 1996. *Beyond intellectual property:Toward traditional resource rights for indigenous peoples and local communities*. Ottawa etc.: International Development Research Centre.

Pottier, J. 1993. Harvesting words? Thoughts on agricultural extension and knowledge ownership, with reference to Rwanda. *Entwicklungsethnologie* 2(2): 28–38.

Prattis, J.I. 1973. Strategizing man. *Man* 8(1): 45–58.

Quarles van Ufford, P. 1993. Knowledge and ignorance in the practices of development policy. In *An anthropological critique of development. The growth of ignorance*. (ed.) M. Hobart. London, New York: Routledge. 135–160.

Quinn, N. & D. Holland. 1987. Culture and cognition. In *Cultural models in language and thought*. (eds.) D. Holland & N. Quinn. Cambridge: Cambridge University Press. 3–40.

Raethel, A. 1993. Auswertungsmethoden für Repertory Grids. In: *Einführung in die Repertory Grid-Technik. Bd. 1: Grundlagen und Methoden*. (eds.) J.W. Scheer & A. Catina. Bern: Verlag Hans Huber. 41–67.

Richards, P. 1980. Community environmental knowledge in African rural development. In *Indigenous knowledge systems and development*. (eds.) D. Brokhensha, D.M. Warren & O. Werner. Washington, D.C.: University Press of America. 181–194.

Rogoff, B. & Lave, J. (eds.) 1984. *Everyday cognition:Its development in social context*. Cambridge, Mass.: Harvard University Press.

Saegert, S. & G.H. Winkel. 1990. Environmental Psychology. *Annual Review of Psychology* 41: 441–477.

Sankoff, G. 1971. Quantitative analysis of sharing and variability in a cognitive model. *Ethnology* 10: 389–408.

Schank, R.C. & R.P. Abelson. 1977. *Scripts, plans, goals, and understanding: An inquiry into human knowledge structures*. Hillsdale: Erlbaum Publishers.

Scheer, J.W. 1993. Planung und Durchführung von Repertory Grid-Untersuchungen. In: *Einführung in die Repertory Grid-Technik. Band 1: Grundlagen und Methoden*. (eds.) J.W. Scheer & A. Catina/ Bern: Verlag Hans Huber. 24–40.

Scheer, J.W. & Catina, A. (eds.) 1993 *Einführung in die Repertory Grid-Technik. Band 1: Grundlagen und Methoden*. Bern: Verlag Hans Huber.

Schmit, L. 1996. The deployment of civil energy in Indonesia: Assessment of an authentic solution. In: *Civil society: challenging western models*. (eds.) Ch. Hann & E. Dunn. London & New York: Routledge. 178–198.

Schönhuth, M. & U. Kievelitz. 1995. *Participatory learning approaches: Rapid Rural Appraisal / Participatory Appraisal: an introductory guide*. Roßdorf: GTZ.

Scott, J.P. 1998. *Seeing like a state: How certain schemes to improve the human condition have failed*. New Haven & London: Yale University Press.

Seur, H. 1992. The engagement of researcher and local actors in the construction of case studies and research themes: Exploring methods of restudy. In *Battlefields of knowledge: the interlocking of theory and practice in social research and development*. (eds.) N. Long & A. Long. London & New York: Routledge. 115–143.

Shiva, V. 1997. *Biopiracy: The plunder of nature and knowledge.* London: South End Press.
Sillitoe, P. 1998a. The development of indigenous knowledge: a new applied anthropology. *Current Anthropology* 39(2): 232–252.
Sillitoe, P. 1998b, What know natives? Local knowledge in development. *Social Anthropology* 6(2): 203–220,
Sillitoe, P. 1998c, Knowing the land: Soil and land resource evaluation and indigenous knowledge. *Soil Use and Management* 14: 188–193.
Sinclair, F.L., D.H.Walker, B. Thapa, L. Joshi, P. Preechapanya & A.J. Southern. 2000: *General patterns in indigenous knowledge.* Paper given at ASA-conference on 'Participating in Development' April, 2nd to 5th 2000 at SOAS in London.
Siverts, H. 1991. Technology & knowledge among the Jivaro of Peru. In: *The ecology of choice and symbol: Essays in honour of Fredrik Barth.* (eds.) R. Gronhaug, G. Haaland & G. Henriksen. Bergen: Alma Mater Forlag AS. 297–311.
Sperber, D. 1982. *Le savoir des anthropologues.* Paris: Hermann (Collection Savoir).
Spradley, J.P. 1980. *The ethnographic interview.* New York etc.: Holt, Rinehart & Winston.
Stokols, D. & I. Altman. (eds.) 1987. *Handbook of environmental psychology, Vol. 1 und 2.* New York: Wiley.
Strathern, M., M.C.da Cunha, P. Descola, C.A. Alfonso & P. Harvey. 1998. Exploitable knowledge belongs to the creators of it: a debate. *Social Anthropology* 6(1): 109–126.
Strauss, A.L. 1987. *Qualitative analysis for social scientists.* Cambridge: Cambridge University Press.
Strauss, C. & N. Quinn. 1997. *A cognitive theory of cultural meaning.* Cambridge: Cambridge University Press.
Sudarmo, S.P. 1997. Recent developments in the Indonesian urban development strategy. In. *The challenge of sustainable cities: neoliberalism and urban strategies in developing countries.* (eds.) M. Burgess, M. Carmona & T. Kolstee. London: Zed Books. 230–244.
Tambiah, S.J. 1990. *Magic, science, religion, and the scope of rationality.* Cambridge: Cambridge University Press.
Tanner, C. & K. Foppa. 1996. Umweltwahrnehmung, Umweltbewußtsein und Umweltverhalten. In *Umweltsoziologie.* (Hrsg.) A. Diekmann & C.C. Jaeger. Opladen: Westdeutscher Verlag (*Kölner Zeitschrift für Soziologie und Sozialpsychologie*, special issue) 245–271.
Thrift, N. 1996. Flies and germs: A geography of knowledge. In: *Spatial Formations.* N. Thrift. London: Sage. 96–124.
Tirtosudarmo, R. 1997. Economic development, migration, and ethnic conflict in Indonesia: a preliminary observation. *Sojourn. Social Issues in Southeast Asia* 12(2): 293–328.
Tognoli, J. 1987. Residential environments. In. *Handbook of environmental psychology, Vol. 1 und 2.* (eds.) D. Stokols & I. Altman. New York: Wiley. 655–690.
Van der Ploeg, J. 1989. Knowledge systems, metaphor and interface: the case of potatoes in the Peruvian highlands. In: *Encounters at the interface: a perspective od social discontinuities in rural development.* (ed.) N. Long. Wageningen: Agricultural University (Wageningse Sociologische Studies, 27).
von Cranach, M. 1995. Über das Wissen sozialer Systeme. In *Psychologie des Sozialen: Repräsentationen in Wissen und Sprache.* (Hrsg.) U. Flick Reinbek bei Hamburg: Rowohlt Taschenbuch Verlag. 22–53.
Vayda, A.P. 1983. Progressive contextualization. Methods for research in human ecology. *Human Ecology* 11: 265–281.
Warren, D.M., L.J. Slikkerveer & D. Brokhensha. (eds.) 1995. *The cultural dimension of development:Indigenous knowledge systems.* London: Intermediate Technology Publications (IT Studies in Indigenous Knowledge and Development).
Weber, H. 1994. The Indonesian concept of development and its impact on the process of social transformation. In *Continuity, change and aspirations: Social and cultural life in*

Minahasa, Indonesia. (eds.) H. Buchholt & U. Mai. Singapore: Institute of Southeast Asian Studies (ISEAS). 194–210.

Weller, S.C. & A.K. Romney. 1988. *Systematic data collection.* Newbury Park: Sage Publications (Qualitative Research Methods, 10).

Werner, O. & Fenton, J. 1970. Method and theory in ethnoscience and ethnoepistemology. In *Handbook of Method in Cultural Anthropology.* (eds.) R. Naroll & R. Cohen. New York: The Natural History Press. 537–578.

Werner, O. & M. Schoepfle. 1987. *Systematic fieldwork (2 vols.).* Newbury Park: Sage Publications.

Willutzki, U. & A. Raethel. 1993. Software für Repertory Grids. In: *Einführung in die Repertory Grid-Technik. Band 1: Grundlagen und Methoden,* (eds.) J.W.Scheer, & A. Catina. Bern: Verlag Hans Huber. 68–79.

Chapter 2

Doing and Knowing: Questions about Studies of Local Knowledge

Andrew P. Vayda, Bradley B. Walters and Indah Setyawati

Anthropological claims about indigenous or local knowledge often exaggerate the cultural mystique of such knowledge 'systems' and the difficulties associated with rendering local knowledge accessible to outsiders and with ascertaining its utility for initiatives in economic development and environmental conservation. We argue that part of this confusion reflects a common tendency for researchers to preoccupy themselves with understanding the knowledge *per se* rather than with understanding what knowledge actually influences human behaviour in different situations or contexts and how and why it does so. We propose, as an alternative, an approach which begins by focusing on specific events or actions, rather than knowledge systems, as our objects of study and explanation and then goes on to consider what actors know mainly insofar as that can help us to make sense of the events or actions of interest. Such an approach de-mystifies the notion of knowledge systems and makes productive investigations possible without always requiring laborious investments in ethnographic fieldwork. Illustrations of this approach and of the arguments presented in its support are drawn here mainly from our research in Indonesia and the Philippines.

Like others who are advocates of greater involvement of anthropologists in studies of local knowledge relevant to new initiatives in economic development and environmental conservation (see especially Sillitoe 1998a, 1998c), we are interested in the actions that people take in using and managing their environments or environmental resources, in the knowledge that their taking those actions and not taking certain others is based on, and in the causes of changes in the actions and their knowledge bases. Presumably also shared with these advocates is our belief that studies of these matters can be important for the success of the initiatives in economic development and environmental conservation. Our participation in the studies to be described below was motivated, inter alia, by this belief.

However, we part company with at least some of the advocates on certain methodological issues related to practically relevant studies of doing and knowing. These are the issues that will be addressed here, and we will cite our experience in research projects in Indonesia (Vayda & Setyawati) and the Philippines (Walters) to illustrate how our methods for dealing with the questions differ in practically advantageous and consequential ways from methods that have been employed or advocated in studies of indigenous or local knowledge.[1]

In brief, our view is that anthropologists can deal more effectively and expeditiously with the matters specified in the first paragraph if they do not commit themselves to so-called holistic studies of necessarily shared or socially or culturally embedded local knowledge. In support of this view, we will present methodological arguments echoing those we have previously made elsewhere (especially in Vayda & Walters 1999; see also Vayda 1996, 1997, 1998, Vayda & Sahur 1996, and Vayda & Setyawati 1998). On the one hand, these are arguments in favour of being guided in our research by questions about the causes of outcomes of interest. On the other hand, they are arguments against limiting those questions to how-questions, concerned with how factors privileged in advance by the investigator influence such outcomes as environmental changes and the actions that people take in using and managing their environmental resources. These arguments were made in some of our previous publications (especially in Vayda & Walters, 1999) against privileging certain kinds of political factors in advance, but here we will make them against so privileging cultural knowledge and so-called indigenous knowledge systems. And we will further argue here that eschewing such privileging is important for making our studies bear more effectively and expeditiously on development and conservation initiatives.

Our arguments for studying local practice and knowledge without necessarily engaging in long-term, holistic ethnographic research should, however, not be construed as opposition to all in-depth, ethnographic fieldwork in relation to development and conservation initiatives. In fact, as the case studies in the following two sections should make clear, the amount of time and effort invested in fieldwork will need to vary, depending on a variety of pragmatic considerations. Clearly there will be cases where in-depth ethnographic studies are needed. But the fact that the results from such studies *sometimes* justify the time, effort, and expense put into them cannot be used to argue that the studies must *always* be undertaken. By the same kind of illogic, the fact that the studies are sometimes unproductive could just as well be used to argue that they should *never* be undertaken.[2] There have indeed been unproductive studies, i.e., research leading to the discovery of local knowledge and practice that turn out on analysis to be 'useless, harmful, or otherwise scientifically indefensible' (Lees 2000, citing Hess 1997). But, as Hess (1997: 79) has noted, such cases are, by and large, omitted from the literature on local or indigenous knowledge systems. In her report on her own research on local knowledge and practice related to sheep management in an Andean community in Ecuador, Hess does ask for granting the same validity to our own and indigenous views of reality, including the different views of the etiology of disease in sheep. However, despite this appeal, she brings forth no local knowledge and practice on which development initiatives might build. On the contrary, her account indicates that local knowledge and practice have resulted in sick and unproductive sheep and impoverished people.

As has already been stated, it would be illogical to use such cases to argue against ever undertaking in-depth, holistic studies of local knowledge. The cases do, however, support our view about the need for making local knowledge studies bear more effectively and expeditiously on development concerns and initiatives. In light of the possibility that there will be no significant payoff from some such studies (regardless of how much time, effort, and resources are invested in them), it becomes all the more important to find and adopt ways of zeroing in on the local knowledge and practice which do have significant and positive practical relevance to development and conservation initiatives.

Our arguments here are in accord with certain points that others have made in favour of 'rapid rural appraisal' and similar shortcut, rapid research methods in development studies. In particular, we agree with points made by Chambers (1991: 522) to the effect that it is important to know what is not worth knowing and to abstain from trying to find it out. Regrettably, however, such points are left virtually as slogans by Chambers and like-minded advocates in the development-studies field. Unlike us, they are not concerned with explicating procedures in research guided by questions about the causes of outcomes of interest. Accordingly, no clear advice is to be found in their writings about how, in the course of such research, decisions are to be made about what is worth knowing and what is not. Thus, notwithstanding all their discussion of surveys, check-lists, flow charts, diagrams, and other methods for obtaining and recording locally available information (see, for example, Chambers 1991, Mikkelsen 1994), what is obtained and recorded by their methods is, both in our view and in that of some other critics (e.g., Pelkey 1995), too often only background information for the more sharply focused inquiries needed to produce usable evidence for or against particular causal possibilities in the kind of research that we are advocating (cf. Walters et al. 1999: 209–210).

Our own research procedures will be explicated mainly in the context of the case studies presented in the next two sections. The following key features of our approach are, however, worth noting here:

1. Identifying at the outset certain environment-related or resource-related actions as objects of study on the basis of their relevance to development and/or conservation concerns or programs rather than on the basis of their meeting conventional criteria for anthropological subject matter.

2. *Not* undertaking studies of knowledge per se and *not* singling out shared or so-called cultural knowledge for investigation but, instead, trying only to ascertain any local knowledge likely to be useful to us for deciding about focusing on and explaining particular actions because of their relevance to development and/or conservation concerns.[3]

3. *Not* assuming that the practices and knowledge behind them which are of interest to us are embedded in a 'whole system' (Sillitoe 1998a: 247) or 'encompassing cultural matrix' (Ellen 1998:238) which must be elucidated or comprehended in toto if we are to understand the practices and knowledge well enough to use them effectively to meet development and/or conservation goals. Instead, subscribing on these matters to the views of such philosophers as Lewis (1986) and Rorty (e.g., 1989: chap. 1) and such social scientists as Hawthorn (1991:173–174, 180, *passim*), we assume the following:

a. That understanding or explanation of anything that people do or know can be based on seeing or showing its connections to any number of other things or events, whether within an encompassing cultural matrix or not.

b. That partial explanations, indicating only some connections and missing others, are useful and, practically speaking, necessary.[4]

c. That our decisions about which connections to pay more or earlier attention to may be made on pragmatic grounds, such as those discussed in the next two sections and in our earlier publications (e.g., Vayda 1983 and 1996:9–16, 22–24), rather than on the basis of theories or discipline-rooted biases about the kinds of systems within which connections with explanatory import must be sought.

Before proceeding to the case studies, we must make clear that we have been referring to *cultural embeddedness* only because of the importance assigned to it by others presenting anthropological views of indigenous or local knowledge (e.g., Sillitoe 1998a; Ellen 1998). Because we share the view of some critics (e.g., Sunley 1996: 345–346; Portes & Sensenbrenner 1993) who argue that concepts of cultural or social embeddedness generally suffer from vagueness and that their significance or utility remains uncertain, we do not employ any concept of embeddedness when we report on our own research.

There is, however, something to be said about a special, relatively clear sense in which a concept of embeddedness is sometimes used. In this sense, embedded knowledge refers to knowledge that local people have gained from their forbears about ways of doing things but without knowledge of why those ways work. What is known is simply that those ways must be used if the crops are to grow well or if the prey is to be captured (see, for example, the discussion in Alcorn 1989: 65–66). Although not encountered in the course of the Indonesian and Philippine research from which our main illustrations have been drawn for this article, such knowledge might still seem to some anthropologists to need to be considered here. This is because it might seem to them that, on the one hand, our less opaque examples of knowledge from Indonesia and the Philippines have led us to overstate our arguments against committing ourselves to holistic studies of cultural systems and that, on the other hand, such holistic studies are still needed to elucidate local knowledge which is embedded in the special sense indicated in this paragraph. Accordingly, an example of such knowledge from the New Guinea highlands will be briefly considered after the Indonesian and Philippine case studies presented in the next two sections.

Studies of knowledge and practice related to insect pest management in Indonesia

As part of a nationwide Indonesian program intended to train 2.5 million rice-growers in integrated pest management (IPM), an invitation was extended at the beginning of 1990 to Vayda and a number of University of Indonesia faculty members and students, including Setyawati, to engage in research on variation and change in agricultural pest management as practiced by Central Javanese rice farmers. The questions which entomologists and other scientists in the program put to us as anthropologists concerned not only the farmers' actual practices and the ideas on which these were based, but also the likelihood of change in the practices with the acquisition of new knowledge such as the IPM training was intended to help develop. Examples of this included knowledge about interactions between insect pests and their predators, about damage caused by different pests at different stages in the rice plants' growth, and about the dangers of resurgence of pests following indiscriminate use of broad-spectrum pesticides which do a better job of eliminating the pests' natural enemies than eliminating the pests themselves. The hope was that farmers who had presumably been overusing pesticides would become ecologically informed decision-makers and actors in their rice-fields. This, suggested one of the entomologists, would mean a 'paradigm shift' among the farmers.

Coming to the project as ecological anthropologists but regarding as its essential subject matter the interrelations of knowledge and action or of cognitive and agro-ecological change, we sought to make up for our own lack of training in cognitive anthropology by reviewing potentially relevant literature in that area. The discussion in this section, incorporating much of a previously published article (Vayda & Setyawati 1998), will be concerned with why that review was disappointing to us and how it led us to see that anthropologists' assumptions about their proper subject matter were keeping them from making their studies more effectively and expeditiously relevant to programs of economic development or environmental management. There will also be further explication of the research procedures that we favour for achieving such relevance.

That much of cognitive anthropology has been little concerned with practical activities or behaviour is well known.[5] Even so, we hoped to find more than we actually did concerning possible knowledge bases of pest management practices. Thinking, for example, that such bases might include specific knowledge of insect reproduction and behaviour, including predator-prey interactions, we hoped to find some good descriptions of people's knowledge, as well as identification of their ignorance, concerning such matters. Instead we experienced disappointment similar to that reported by Bentley, another anthropologist involved in IPM-related research. What he found was that too many accounts ostensibly treating indigenous knowledge of the natural world emphasized linguistic distinctions with little practical relevance. He also found that even a work as detailed and ethnographically far-ranging as Hunn's on zoological classifications by Tzeltal-speaking Indians in Chiapas (1977) was deficient in describing those aspects of the people's knowledge and ignorance which could be important for our understanding of their actual or potential pest management practices (Bentley & Andrews 1991: 117).[6]

Our disappointment did not, however, end with accounts in which behaviour was, at best, a secondary concern. It extended to studies and methodological prescriptions by some in cognitive anthropology who were vocal in decrying inattention to behaviour. The problem for us here was somewhat different from the already mentioned problems with studying knowledge prior to ascertaining that it lies behind some actions or behaviour of actual or potential practical significance. The problem here lay in the restrictive nature of these cognitive anthropologists' definition of both their behavioural and cognitive subject matter. In line with Goodenough's original mandate (1957) for cognitive anthropology to study whatever one must know to behave in a culturally appropriate or societally acceptable manner, the knowledge featured in their studies and programmatic statements was shared, so-called cultural knowledge, while the featured behaviour was behaviour that is (or could be presumed to be) culturally appropriate or societally acceptable or, in some important way, influenced by culture-specific models of the world (see, for example, Quinn & Holland 1987; Keller & Keller

1993; Goodenough 1994). Even when 'ecological effectiveness' was explicitly a concern of the analyst, studying how such effectiveness is achieved through the pursuit of culturally defined goals was the program recommended (Hunn 1989: 145).

We realized in due course that such programs were at odds with our desire, supported by our entomological colleagues, to be expeditious in making our studies relevant to Indonesia's IPM program. In light of this, our bases for deciding what actions and knowledge to pay attention to and investigate had to be different from the culture-related considerations used by cognitive anthropologists whose work we had reviewed. Our criterion had to be whether the behaviour and its knowledge bases were, or might be, significant for (or as) actual or potential pest management, regardless of what we could see or hope to see about its being shared, culturally appropriate, culturally influenced, or societally acceptable. This criterion was geared to our project's practical goal of change in pest management by Indonesian farmers. However, the criterion also was compatible with more academic goals which we wished to pursue in the project, for example, the goal of further developing research procedures and explanatory models concerning interrelations of actions, the actors' reasons for them, and the contexts in which they occur. As indicated elsewhere (e.g., Vayda 1983, 1994, 1996, 1998; Vayda & Walters 1999) and as, in effect, is illustrated in the present article, the actions to which these procedures and models apply are whatever actions occur rather than just the kinds of standardized actions on which some cognitive anthropologists have been wont to focus.

Having said all this, we also have to make clear that in the IPM project we were simply not privileging rather than shunning more traditionally anthropological lines of inquiry (cf. Barth 1994: 358 on focusing on acts and events rather than privileging 'culture'). Certainly we were not uninterested in the influence of shared or cultural knowledge and beliefs on the behaviour we were studying. Indeed, we were well aware that some of the novel or idiosyncratic pest management practices and ideas that we were finding might constitute culture in the making. However, the fact remains that behaviour did not have to pass some kind of cultural test in order to become the focus of our investigations and analyses.

What then was some of the behaviour on which we concentrated our attention? What knowledge and ideas as bases for the behaviour did we find among the farmers? What relevance to the IPM program did our findings have? And are there more widely applicable lessons here for conducting practically relevant studies of interrelations of knowledge and action and for making distinctively anthropological contributions to such studies? These are some of the questions to which the remainder of this section will be devoted.

Our examples of behaviour are drawn mainly from five villages in which Setyawati and four of her fellow students from the University of Indonesia conducted field research for seven months during 1990.[7] The villages are located in the four regencies of the Special Region of Yogyakarta. Weekly morning-long IPM training sessions, including both field and classroom exercises, were held for approximately 25 farmers in each of the villages for ten weeks. Each farmer received 1,000 rupiahs (about US$0.55 in 1990) and a snack for each session attended. Research involved observation of these sessions (the 'field schools'), as well as other participant observation and interviewing in the villages.

Behaviour learned in the field schools was certainly interesting to us, and, for making assessments of needs and prospects for IPM, we were especially interested in seeing whether or how or to what extent such behaviour was continued outside the schools (cf. Lave 1988: chap.2 and *passim* on learning-transfer research). An example concerns making observations of the numbers of insect pests and predators in the fields in order to provide a basis for ecologically informed decisions about pest management. The specific procedures taught in the schools for making such observations involved counting insect pests and their predators in sample rice plants along a transect in a rice-field in order to decide whether to spray the field with a chemical pesticide. Of interest to us were not only these procedures but also any alternatives to them developed by the farmers themselves.

We found that many of the farmers succeeded in learning the prescribed sampling and counting procedures and in putting them into practice in the fields rented for use in the school sessions. However, although the farmers had been told in the schools to follow the same procedures in the fields in which they were growing their own crops, we did not find any farmers doing this. Going with the farmers to their fields, we sought therefore to identify what they did instead to decide whether to spray or not. We found some walking through their fields, usually near the edges, and looking out for the more visible pests such as grasshoppers and rice seed bugs (*Leptocorisa*). Sometimes, as they walked, they also stretched their arms out sideways and passed them back and forth over the tops of the rice plants in order to see how many or how dense were the stemborers (*Scirpophaga*), rice seed bugs, and possibly other insect pests flying forth. Mostly the farmers did not walk into the fields at all but instead inspected them for pests from the bunds. These inspections, whether by walking through the fields or only along the bunds, were for the pests alone and not for their predators.

The observed differences between the farmers' practice in the schools and outside them led to interviews which indicated to us that evidence of whether the paradigm shift hoped for in the IPM program was being achieved had to consist of more than seeing whether the procedures of Western scientists were being copied. As already noted, there indeed were farmers who could and did learn sampling rules and counting procedures; however, they regarded these simply as rules and procedures to be followed in the schools and they saw no good reason for applying them to their own farming activities. Neither the logic behind sampling, nor predator-prey dynamics among insects (rather than among larger creatures), were understood well enough by the farmers in 1990 for them to devote – or probably to even consider devoting – the extra time and effort which would have been required to monitor the fields as prescribed at the time by the IPM program. That they had in fact decreased their use of pesticides was explained by some farmers by saying that pesticides had become too expensive. The rise in pesticide prices had occurred with the gradual elimination of subsidies after IPM had become the national pest management strategy by presidential decree in 1986.

We were able to return to the study area for only a brief stay in 1992, i.e., just long enough for single-day visits to two villages where field schools had been in operation for two years. These visits sufficed, however, to impress upon us that some farmers attending the schools had become quite knowledgeable about predator-prey interactions affecting insect pests. They told us that their knowledge of these insect interactions had been gained in the schools, although some referred as well to their pre-school knowledge of *non-insect* predator-prey relations, including knowledge of predation by snakes on rodent pests in rice-fields (cf. van de Fliert 1993: 97).[8] The newly knowledgeable farmers told us too that they had cut their use of pesticides because they were making their decisions about pest management on the basis of their observation of numbers of both insect pests and predators in their fields. This impressed us as being in accord with the paradigm shift hoped for by our entomological colleagues. However, the farmers were not following the sampling and counting procedures prescribed in the field schools at the time of our original research. In fact, after we had made our observations of differences between what farmers did in the schools and out of them in 1990, IPM training had been modified so that it was, in 1992, regarded sufficient for farmers, as distinct from IPM researchers, to make reasonably accurate estimates rather than exact counts (Gallagher n.d.). One farmer that we talked to had, on his own, hit upon the technique of monitoring stemborers by means of seeing what densities of stemborer moths were attracted to lights that he flashed in his field at night. It was by our criterion of practical relevance that such methods of estimating pests and predators clearly merited our attention, but, as will be discussed shortly, the methods are anthropologically interesting as well because of the questions they raise about alternatives to their being interpreted or explained as culture-specific practices or developments.

During our original research, the other behaviour to which we paid attention included, at least initially, whatever farmers did or had done with pest management as their goal. Our thinking on this was that, once any such practices had been identified, we would look at them more closely if it seemed to us and to our entomological colleagues that the practices could actually achieve, or be adapted to achieve, the pest management intended by the farmers. One example of such a practice is using crabs impaled on sticks, which were planted at the edge of the rice-fields, to attract rice seed bugs so that they could be set fire to. Among other examples are practices that are the same as those recommended by Western scientists, e.g., flooding rice-fields to drown mole crickets (*Gryllotalpa*) and draining the fields to rid them of nematodes (see the recommendations in Grist 1986: 345–347). As already noted, it did not matter to us whether these and other practices that we were finding, and whatever knowledge we were finding them to be based on, were already widely shared or evidently culturally embedded, or whether on the contrary, they were novel or idiosyncratic. If we were to find the latter to be the case with respect to something actually or potentially effective for pest management, we hoped we could proceed, in follow-up research, first to discover factors (including cognitive ones) contributing to its being practiced by some and not by others and then to assess the possibilities for its being more widely adopted. As for old and disused practices that were still remembered, we were interested in pursuing inquiries about those too but, in line with our criterion of practical relevance, not if they clearly held no promise of being revived and/or being efficacious for pest management in the future. An example of such an unpromising practice, described to us by a few informants, consists of catching a certain uncommon species of grasshopper in the bush or forest and then carrying it around the rice-fields and chanting to the rice seed bugs: 'This is your father [or, according to one informant, your husband]; go with him; your home is not here.'

Also of interest to us were certain instances of deliberately not taking action or ceasing to take action against pests. When not doing something is deliberate, it may call for explanation just as much as doing something does.[9] In our particular project, it was relevant for pragmatic reasons to ascertain whether inaction might have causes that could keep farmers from being or becoming ecologically informed pest managers. The cases in question concerned certain armyworm (*Spodoptera*) and brown planthopper (*Nilapavarta lugens*) attacks that farmers told us about. They said that, in the early stages of rice-field infestation, they acted on the basis of practical ideas about chemical control of the pests. However, as infestation became more severe and the measures taken proved unavailing, they decided they could do nothing about it. Some said that they resigned themselves to loss of their crop and sought temporary employment as pedicab-drivers or construction workers in the city of Yogyakarta. Reasons for inaction were found by some farmers in traditional knowledge of the supernatural. To these farmers, the failure of insecticides in these cases was proof of two things: i) that the pests in question had been magically brought forth from her realm by Nyai Loro Kidul, the goddess of the Southern Ocean; and ii) that the pests would not leave before sating themselves on the crop. Indeed, to some farmers, any major pest outbreaks were attributable to Nyai Loro Kidul.

With respect, however, to the brown planthopper outbreaks, other farmers drew on other knowledge to contest this view. Specifically, they argued against it by noting the following: i) the absence of such outbreaks in former times when the pests that the goddess sent forth were rats (cf. Becht 1939); ii) the fact that the planthoppers did succumb sometimes to pesticides; and iii) the fact that visitations by them, unlike those by armyworms, did not occur suddenly and as if by magic but rather were preceded by such warning signs as strong winds or heavy rain. In the view of some farmers, planthoppers were actually brought from the north by wind and rain. Our conclusion from these cases, as well as our overall impression from observations made in the course of the project, was that mystical knowledge or magical beliefs did not keep farmers from putting practical measures for pest control into effect. Such knowledge or beliefs were invoked by some farmers to justify inaction when the practical measures that they had taken did not work, while there were other farmers who found reasons not to use the same knowledge or beliefs even in these circumstances. As one of us has suggested elsewhere (Vayda 1996: 6), recourse to such ideas at some times and not others by the same farmers dealing with the same pests shows that we need to pay attention to context-dependent variability in whether or not particular knowledge or ideas or cognitive attitudes are efficacious in guiding actions (cf. Bratman 1992; Thomason 1987). In other words, there is an argument here not only for focusing first on actions rather than studying knowledge per se but also for focusing, as we had, on the contexts in which the actions occur in order to come closer to pinpointing the knowledge on which the actions are based.

A more general argument may be set forth here as well in order to close this section. It is worth noting that once that the criterion of actual or potential relevance to pest management had been used to make certain actions our objects of study in the IPM project, we could proceed in what Barth (1987: 24) has described as 'the tradition of the wondering naturalist.' That is to say, we could look closely at particular situations in the search for whatever factors may have been operating to produce those actions, without *a priori* theory-based or discipline-rooted constraints on the factors we could consider or accord priority to.[10] Thus, when we focused on cultural factors (whether Javanese mystical knowledge and magical beliefs or more practical cultural knowledge), it was not because such factors are what anthropologists *must* focus on. Rather it was because it seemed to us, at particular times in the course of our research, that particular cultural factors were likely to have affected behaviour we needed to explain. For example, when it was found that some farmers had gained impressive knowledge of insect predator-prey relations and were applying it by making field observations of insect pests and predators to provide a basis for pest management decisions, our attention was drawn to the fact of widely shared pre-existing knowledge concerning *non-insect* predator-prey relations, including the knowledge that many farmers had of predation by snakes on rodent pests in rice-fields. We proceeded then to ask whether farmers were drawing on this knowledge in making their decisions to commit themselves to field observations of insect pests and predators.

However, at other times in the course of the research, we were just as ready to focus on factors other than culture-specific ones to explain behaviour. A case in point is the farmers' making only visual estimates of insect densities and usually doing so by inspecting fields from the edges rather than following prescribed sampling or counting procedures. We recognized this to be like behaviour observed by others in similar situations outside Java (e.g., in Honduras as described by Bentley and Andrews 1991: 118 and in the Philippines as described by Kenmore et al. 1987: 106). Accordingly we sought to understand and explain the behaviour less as the product of culture-specific influences than as the product of such factors as time constraints, lack of practical experience of sampling, and habits of reliance on frequency estimates and statistical judgments like those which, according to some psychologists (e.g., Gigerenzer 1991, 1992; Cosmides & Tooby 1996), are intuitively made by people everywhere and are often adequate for the objectives at hand. Anthropologists intent on making their work as practically relevant as possible will need sometimes to pay close attention to such factors instead of feeling that they must always concentrate on factors to which they are directed by anthropological theories or by norms about anthropological subject matter.

Studies of knowledge and practice related to mangrove tree planting in the Philippines

Illustrations from the Philippines are drawn from field research by Walters in 1997 on the causes and consequences of mangrove tree planting in Bais Bay, Negros Oriental (Walters 1998, 2000a, 2000b). As in the IPM research, investigations of mangrove planting led to the finding that practically relevant knowledge was often not widely shared in the community. Specific to this case, however, was the added discovery that knowledge factors were, in general, not always important for our understanding or explaining actions or behaviours of interest.

Mangrove forests provide important ecological services and are commonly exploited by coastal communities for firewood, construction materials, and various fish, crustaceans, and shellfish (Dewalt, Vergne & Hardin 1996; Diop 1993; Hamilton, Dixon & Miller 1989; Kunstadter, Bird & Sabhasri 1986; Lacerda 1993; Macnae 1968; FAO 1994). Seventy percent of mangroves in the Philippines have been cleared since 1940 to make way for brackish water aquaculture, residential settlement, and various public and commercial infrastructure developments (Baconguis, Cabahug & Alonzo-Pasicolon 1990; Primavera 1995). Moreover, most remaining forests are degraded as a result of past and continued wood harvesting. This state of affairs has prompted a variety of policies and programs to protect and restore existing mangrove sites (Calumpong 1994; DENR 1990; Walters 1995). In particular, the deliberate planting of mangroves is now enthusiastically promoted by governments, NGOs, and aid agencies in the Philippines and elsewhere as a means to restore degraded ecosystems while enhancing livelihood options for poverty-stricken, coastal communities (e.g., Kaly & Jones 1998; Lewis 1990; Pomeroy et al. 1996; Primavera & Agbayani 1996; Thorhaug 1990; Sukardjo & Yamada 1992; Saenger & Siddiqi 1993; Van Speybroeck 1992).

This broader context made significant the discovery of cases where local fisherfolk and fishpond owners have been planting and managing mangrove forests for decades under their own initiatives (e.g., Yao & Nanagas 1984; Cabahug et al. 1986; Walters 1995, 1997).[11] The reasons for planting are varied and include the desire to provide a ready supply of construction materials and firewood, to increase tenure security, and to protect seaside homes and fishpond dikes from storm damage (Walters 1997, 2000a, 2000b). The extent to which these existing management systems might serve as a model or offer lessons to wider programs in mangrove restoration was an important consideration for Walters in undertaking the study. Accordingly, he sought specifically to understand the origins and spread of this 'indigenous' planting. The possible importance of knowledge factors in its spread was recognized from the outset, and such factors were given due attention in the course of the research whenever they were seen to be relevant to explaining mangrove planting.

Investigations led Walters to discover, first, that planters are many and diverse in the few villages where mangrove planting does occur. It is typically a private, household-based activity, but participation does not otherwise strongly correlate with education, income, or other socio-economic characteristics. The ubiquity of mangrove planting in these sites led Walters initially to speculate that there must be a well-developed, widespread and systematic knowledge base for planting in communities where it is common. However, investigations in villages in Bais Bay, where most of the fieldwork was done, revealed much variation in knowledge between planters and little evidence of widely held, systematic knowledge about mangroves and mangrove planting. In fact, many planters in Bais did not give much thought to their planting activities. Often, when made impatient or annoyed by questions intended to elicit their knowledge of the subject, they insisted that planting is 'easy, you just get the mangrove and stick it in the mud and it grows!'

That these persons were able to plant at all reflects the fact that the mangrove planting is indeed not a basically complex task and requires little technical knowledge to accomplish. More specifically, it may be noted that *Rhizophora* species are the preferred and most widely planted mangrove trees (Walters, 2000b). Rhizophora are viviparous, meaning that seeds begin to grow and elongate into stems while still attached to the parent plant (Tomlinson 1986). To plant, one simply collects the elongated seedlings, called 'tawin,' from the parent tree when ripe – a condition easy to assess – and places them $\frac{1}{4}$ – $\frac{1}{3}$ their length deep in mud. If environmental conditions are suitable and the young plants are not damaged by disturbance (see below), they will sprout leaves and grow.

These very basic planting techniques are straightforward and widely practiced in Bais. However, planters were found to vary considerably in terms of the deployment of other, more specific practices. Such variation was found, in cases, to reflect genuine differences in knowledge between planters and, either demonstrably or else possibly, to have important consequences for planting success. The remaining discussion in this section will concern three such practices: spacing, a method of controlling shell infestation, and test planting.

Something shown to have important consequences for subsequent survival and growth of plantation trees is the initial spacing distance applied between planted seedlings (Shepherd 1986; Wadsworth 1997). In Bais, 47% of the planters in Walters's sample (n=91) used 30–60cm spacing, and 41% used spacing of only 10–20cm. The more experienced and knowledgeable planters almost always employed wider spacing (30–60cm or more) with a clear understanding of its benefits, i.e., compensation for a degree of post-planting mortality and facilitation of the rapid growth of straight stems, most of which are subsequently harvested for fish-corral construction.[12] For example, experienced planters tend to use closer spacing (30–40cm) in areas where they anticipate, based on prior planting experience, relatively high post-planting mortality as a result of wave damage or other disturbance (see below). In sites where post-planting mortality is anticipated to be minimal, wider spacing (40–60cm) is typically used.

Other factors were sometimes found to be taken into consideration by planters when deciding on the appropriate spacing to use. For example, knowledge of local fishing, shell collecting, or boat passage practices led some planters to increase their spacing distances so as to enable those engaging in the practices to pass through planted areas without damaging young seedlings. Some planters were also found to initially use wider spacing (around 1.0m) so that they might quickly establish claims over planting sites. In such cases, these same planters might later return and inter-plant to establish more desirable, higher densities.

What of the 41% of planters in Bais who use spacing of only 10–20cm? Spacing this close is costly in terms of added planting expense (halving the spacing distance quadruples the number of seedlings required). It also creates extremely high stocking densities that, in the absence of very high seedling mortality, almost certainly lead to crowding, reduced stem growth and subsequently high tree mortality (Shepherd 1986; Wadsworth 1997). Interviews of planters who used very close spacing like this confirmed Walters's suspicion that such practices were often ill-advised. Planters who used very close spacing were, with few exceptions, less experienced and often had small plantations.[13] When asked to talk about the very close spacing they used, they often displayed a lack of awareness of the relationship between spacing and growth and/or they cited reasons for using close spacing that were suspect. For example, a number of planters explained that they had to compensate for having such a small area by planting more trees on that area! In short, many had planted without giving much thought at all as to the consequences of specific spacing practices, a fact less surprising given that most of these planters fish for a living and so lack appreciable horticultural experience.

A second practice of interest, the application of used engine oil to control shell infestations, came to Walters's attention during investigation of a particular site in Bais where only one person had managed to plant mangroves successfully. Upon questioning residents in the area, Walters was surprised to learn that many in this particular area had also planted mangroves, but these had all been killed by shell infestations.[14] Further inquiry revealed that the lone, successful planter had repeatedly brushed used engine oil onto the stems of the young plants in order to prevent such infestations. The planter said he had learned this technique during a casual conversation with a man whose house he had been hired to build some years previously and who, by coincidence, was a teacher at the nearby fisheries college and a former project manager for an aid-financed mangrove reforestation project. In his view, the technique was effective by virtue of making stem surfaces slippery and thus keeping young barnacles and oyster shells from attaching. However, the oil may also have some toxic effect whereby young shells and barnacles are killed directly and/or barnacle larvae are repelled from the stem surfaces. We are not aware of any rigorous tests of the efficacy of the technique.

It is nevertheless interesting that interviews with the successful planter's neighbours revealed that none were aware of his method for addressing the problem of shell infestation. Even though they had wondered about their neighbour's success in planting, they had not actually asked him how he had achieved it while they had failed. Instead, they simply gave up, citing explanations of varying degrees of plausibility for their failures. The successful planter likewise had not been motivated to offer to share his trade secret with his neighbours; he cited *eja-ejas* ('each to his own') as the prevailing ethos in such affairs. More will be said about this later.

The third practice of interest was the frequent use of test-planting, in which persons plant prospective sites with small numbers of seedlings in order to evaluate the suitability of those sites for more extensive, subsequent planting. Test-planting was found to be commonly employed by the more ambitious, entrepreneurial, and typically prolific planters. These persons know that mangroves will grow only within a relatively narrow range of environmental conditions, i.e., between the mean- and high-tide levels (Macnae 1968). But, based on their experience as fishermen and planters, they also know that intertidal lands are changing remarkably quickly in some parts of Bais Bay as a result of fishpond development, sediment deposition from rivers, and natural colonization and planting of mangroves (Walters 2000a).[15] Such processes, sometimes creating and sometimes eliminating habitats suitable for mangroves, change local opportunities for planting.

As well, planters in Bais face an onslaught of environmental events, including storms, shell infestations, entanglement by floating seaweeds, and anthropogenic disturbance,[16] that can quickly destroy a young plantation (Vayda & Walters 1999; Walters 2000a). Experienced planters recognize that success depends to a large degree on chance events (e.g., storms and certain anthropogenic disturbances) and factors that vary greatly over space and change quickly in time (e.g., shell infestation, sedimentation) and so are difficult to predict. They thus treat each planting as a distinct, trial and error experiment with the objective of evaluating site suitability. They similarly will observe the plantings of others as tests, even though they rarely actively share what they learn with one another, presumably because good planting sites are rare and an underlying objective of much test-planting is to find these sites before others do. Unlike many of the less experienced or less ambitious planters, these entrepreneurs are not easily discouraged by failure. In some cases, thinking another test is called for, they will replant at a later date; in other cases, efforts at particular sites are chalked up to experience and the planters try elsewhere. As noted above, these planters may vary the spacing or the depth of sowing used, either just to see what happens or because they believe that closer spacing and deeper placement may improve seedling survival in the face of certain forms of disturbance (e.g., waves, floating seaweed, rambunctious dogs). Assessing environmental conditions remains, however, at the forefront of most test-planting efforts.

In summary then, Walters's research led to findings about local knowledge that may be of value for mangrove restoration efforts elsewhere. Thus he learned that basic planting practices are straightforward and easy to learn and apply and so should be easy to introduce into novel settings. However, that these practices are being applied in a given community does not mean that there also is a well-developed, widespread and systematic knowledge base there for planting. In fact, it was found that many planters gave little thought to what they were doing and, as a result, demonstrably or possibly important techniques, such as optimal spacing and using waste oil to control shell infestation, were not widely practised. Walters also learned that environmental events of various kinds often preclude successful planting, regardless of the level of technical knowledge or expertise of the planters. Considerable spatial and temporal variability in the occurrence of such events make it difficult to predict where and when they will constrain planting.

In addition to leading to technical insights about planting, the research also led to valuable discoveries about contextual factors and social processes that influence the spread of knowledge. It was found, for example, that active sharing of knowledge was uncommon among planters in Bais and that, while this partly reflected the often large and amorphous character of villages there,[17] it was also fostered by the competitive nature of planting in a context where suitable planting sites are scarce. In Bais, knowledge of planting is learned by observing one's kin and neighbours, and through direct planting experience (Walters 2000a). The more knowledgeable planters are typically opportunistic and curious in their approach to planting: most of what they know has been learned from ongoing trial and error and from observing others from a distance. However, with so little active knowledge-sharing and so few opportunities to learn directly from experience (because good planting sites are scarce and because, being fisherfolk, most planters have little prior experience of growing other kinds of trees), it is not surprising that many do their planting without drawing on a substantial knowledge base for doing it. Instead they simply observe and copy the basic planting techniques of kin or neighbours. In line with this, one finds little evidence of a complex and widely shared knowledge system to guide the planting of mangroves in Bais. This is consonant also with the further fact that successful planting depends, to a large degree, on environmental events largely out of the control of individual planters.

In fact, the growing realization that knowledge factors were of only limited relevance to explaining the origins and spread of planting led Walters during the course of his research to enlarge the investigation to consider the influence of other factors already alluded to, including the role of environmental constraints and tenure (Walters 1998, 2000; Vayda & Walters 1999). Only about one-quarter of the nine consecutive months spent in the field was ultimately devoted to investigations of knowledge-related matters. The decision to thus extend investigations beyond knowledge as a subject reflected an appraisal of the relative importance of various factors and also the pragmatic consideration that the factors besides knowledge found to be important for explaining planting success and failure in Bais would need to be taken into account in mangrove planting projects elsewhere.

A study of local knowledge in the New Guinea highlands

Is there more of an *a priori* case to be made for holistic studies of cultural systems when knowledge, unlike that in our less opaque Indonesian and Philippine examples, is 'embedded' in the special sense referred to at the end of our introduction, i.e., when it is knowledge which local people have gained from their forbears about ways of doing things but without knowledge of why those ways work? Being confronted with such knowledge may well be a challenge to the researcher, but what we wish to emphasize here is that it is a challenge not necessarily to be met by holistic studies of cultural systems.

A good illustration, cited by one of us elsewhere (Vayda 1995: 227–228) and worth briefly repeating here, is provided by Fringe Enga shifting cultivators in New Guinea's central highlands. Conveyed by them to the geographer Waddell (1972, 1973, 1975) was their knowledge that their sweet potatoes would not grow well unless planted in carefully constructed, large plano-convex mulch mounds, more than a half-meter in height and at least three meters in diameter. Why this was a requirement is something which Waddell's Enga informants were 'not, strictly speaking, able to say' (Waddell 1972: 136). Observations and tests by Waddell (1972: 159–161) and others (e.g., Sillitoe 1998b: 130–132) have indicated that the mounds, by virtue of the mulch in them, contribute to soil fertility (see also the sources cited on this in Sillitoe 1996: 382–383 and 1998b: 130). However, at the time of his field research, it was left to Waddell to surmise that the specific construction and design of mounds at the highest altitudes for growing sweet potatoes could be adaptations to protect the crop from frost. He then proceeded to make observations and tests to confirm this possibility. These further investigations by Waddell, far from being holistic studies of cultural systems, involved, *inter alia*, taking temperature readings from unmounded ground and from the upper parts of mounds and making measurements which showed that the height of mounds and the minimum height above ground at which sweet potatoes were planted increased with altitude, which generally correlated with the intensity and frequency of the frost hazard (Waddell 1973: 36; 1975: 255).

As remarked in Vayda's 1995 article, the thrust of the data obtained by Waddell was to show that the construction and use of mounds were so efficient and precise solutions to the frost problem – so 'well designed,' in other words, to deal with it –

as to make it unlikely that mounding by the Fringe Enga afforded protection from frost just by coincidence rather than as a result of such causes as long-ago trial-and-error learning by some, followed by imitation by others. However, no data were (or could be) obtained on the actual cause-and-effect sequences whereby knowledge of mound construction and of its importance for agricultural success became established or 'embedded' in Fringe Enga culture (Vayda 1995: 228).[18]

There are two points to be made then with an illustration such as Waddell's case. One is that practices supported by opaque professions of knowledge, like those of the Fringe Enga about mounding, may well be important for development, resource management, or conservation and may need to be taken into account in new programs in these areas. The other point is that, with respect to these practices as with respect to some of those considered in the IPM and mangrove-planting cases, studying and explaining the practices so that their practical significance can be recognized and built upon for development or conservation initiatives call for something other than poorly focused inquiries thought to be justified by the mantra of holistic study of cultural systems. In our view, the need is for knowing about and testing situation-specific causal possibilities, such as those related to the New Guinea highland frost hazard as well as those referred to earlier.[19]

Concluding remarks

The foregoing illustrations should suffice as supports for our arguments against using culture-related considerations, including assumptions of embeddedness in cultural matrices or systems, for the purpose, on the one hand, of delimiting the local knowledge and practice to be deemed appropriate for us to study and, on the other hand, of arguing for long-term, in-depth, holistic socio-cultural studies as a prerequisite for recognizing, explaining, and applying local knowledge. Eschewing such considerations for designating what we must study, not only can we still contribute both to practical action programs such as IPM or mangrove restoration and to theoretically meaningful research on the interrelations of cognition and action but also we can, in our view, contribute better and often more expeditiously. We can do this by virtue of being guided in our research more by open questions about why and with what knowledge do people do what they do than by restrictive questions about how actions and knowledge are affected by factors privileged in advance by us because they are cultural and/or assumed by us to be part of (or embedded in) cultural systems that must be elucidated.[20] We can, in other words, do better work and often faster work by not tying our hands in order to make the work distinctively anthropological.

However, as our case studies should have made clear, our saying this does not constitute an endorsement of 'rapid rural appraisal' and similar shortcut, rapid research methods mentioned in our introduction. As stated there, what is obtained and recorded by the methods is too often only background information for the more sharply focused inquiries needed to produce usable evidence for or against particular, situation-specific causal possibilities in the kind of research that we are advocating on the causes of practically relevant actions. We believe that those who have turned to rapid appraisal methods, as much as those still committed to holistic ethnography, can make their research more useful by letting it be guided more, in the ways discussed and illustrated in this article, by clear questions about the causes of concrete actions or events relevant to development and/or conservation concerns.

Acknowledgements

Our thanks to Tom Rudel, Karen O'Neill, Paul Sillitoe, Lisa Gezon, Fergus Sinclair, Paige West, and some anonymous reviewers for comments on earlier drafts of the manuscript.

Notes

1   Sillitoe (1998a: 223, note 2), conceding that it is 'difficult to draw lines between indigenous knowledge, local knowledge, popular knowledge, folk knowledge, and so on', opts for the term 'indigenous knowledge' because it has wider currency in contemporary development discourse than do the other terms (cf. Purcell 1998: 259). All of the aforementioned terms are, however, problematic in one way or another. In this article, we opt for 'local knowledge' because it seems to us a somewhat less incongruous label than 'indigenous knowledge' for some of the recently acquired knowledge of farmers and fisherfolk which is considered in our case studies. As will be apparent, not all of the knowledge to be discussed by us has what some anthropologists engaged in studies of indigenous or local knowledge include among its defining characteristics, i.e., its being clearly local, culturally embedded, time-tested, and intergenerationally transmitted (Berkes 1999: 5–7; Ellen 1998: 238; Hunn 1993).

2   Cf. Huntington 2000: 1273, arguing against either always including or always excluding local knowledge components in environmental management projects.

3   An illustration from the Javanese pest-management studies described in the next section is the action of impaling crabs on sticks and then planting the sticks at the edge of rice-fields. Learning from farmers that they based these actions on their knowledge that the crabs would attract rice seed bugs that could then be set fire to, we looked more closely at the actions and used the farmers' knowledge to help explain them.

4   A classic statement of the complexity of event causation, clearly implying the need for partial explanations, is the following from Carlyle's 1830 essay entitled 'On History': '...actual events are nowise so simply related to each other as parent and offspring are; every single event is the offspring not of one, but of all other events...and will in its turn combine with all others to give birth to new...' (Carlyle 1899: 89). Referring again to the pest-management studies, we may reasonably claim that useful explanations of such actions as the spraying of a particular Javanese rice-field with a particular insecticide at a particular time are possible without our considering all causal antecedents of the actions in chains of events extending back at least as far as the Big Bang.

5   It has been noted both by critics of the field (e.g. Harn's 1968: chap. 20 and others cited in Johnson 1974: 199 and by some of its practitioners (e.g., Gatewood 1985; Johnson 1974; Lave 1988; Quinn and Holland 1987; and, more recently, Hutchins 1995: xi-xii and *passim*; Keller and Keller 1993) and was known to us before we began our literature review.

6   Cf. Last 1981 on the 'reluctance in ethnography to record what people do not know'.

7   The four other students were Rama Chandra, Indrati, Dian Rosdiana, and Bambang Setiawan. We are grateful to them for data that they made available to us. The students

were supervised in the research by Vayda and by Iwan Tjitradjaja and Anto Achadiyat of the University of Indonesia. The program in which our project was included was called 'Training and Development of Integrated Pest Management in Rice-Based Cropping Systems in Indonesia' and was funded mainly by the Food and Agriculture Organization of the United Nations (FAO).

8   Why these farmers previously lacked knowledge of insect predator/prey relations may be related not only to the difficulty of making observations of insect predation with the unaided eye but also to the probable absence of strong incentives for paying special attention to insect pests prior to the first major outbreaks of such pests as brown planthoppers in irrigated rice-fields in the mid-1970s and mid-1980s. These outbreaks were, in large part, a result of inadvertent elimination of the pests' natural enemies by means of sprayed broad-spectrum insecticides, which had only recently been added to Javanese rice farmers' tool-kits as part of the Green Revolution (cf. Kenmore 1991; Gallagher, Kenmore & Sogawa 1994; Settle et al. 1996; van de Fliert 1993: 94).

9   This 'passive counterpart' of action is called 'forbearance' by von Wright (1970: 170–171), but others, like Rheingold (1988: 173–174), have felt that there is a need in English for another word for 'conscious nonaction', Rheingold refers to the term *wei-wu-wei* as meeting the need in Chinese.

10   In line with this are the appeals in Vayda 1998: 578 for 'clear-eyed eclecticism in pursuit of answers to questions about the causes of...concrete actions' and in Vayda and Walters 1999: 171 for not confining our consideration of situation-specific causal possibilities to 'those prescribed by any single or simple agenda or theory'. Also in line with this are arguments set forth a long time ago by the philosopher Charles Peirce (e.g., Peirce 1932: 495–499 and *passim*) and more recently by many others, especially in such fields as artificial intelligence (see, for example, Thagard and Shelley 1997) even if not in anthropology, concerning the explanatorily critical importance of abduction. This, as distinct from deduction and induction, refers to ways of identifying or creating explanatory or causal hypotheses for whatever is to be explained and then deciding how worthy they are of further empirical attention.

11   Local mangrove planting has similarly been documented in Indonesia (Weinstock 1994). Extensive private mangrove plantations in Manila Bay also were described early this century by Brown & Fischer (1918). As far as we know, all of these have since been cleared and the areas developed into fish ponds or reclaimed for residential housing and urban-industrial infrastructure.

12   Foresters typically recommend initial spacing of 1.5 – 3.0m for most upland tree species (Shepherd 1986; Wadsworth 1997) and 1.0m or more for mangroves (e.g., Khoon & Eong 1995; Siddiqi & Khan 1990). Researchers who first documented indigenous mangrove plantations in the Philippines were, in fact, initially perplexed by the widespread use of close spacing (e.g., Cabahug et al. 1986; Emma Melana and Calixto Yao, personal communication). They responded by conducting formal field trials comparing tree growth and survival of planted mangroves at different spacings (0.5x0.5m, 1x1m, 2x2m and 3x3m). In general, these studies show that close spacing of 0.5m does not appreciably reduce growth in the early years and is practical, in particular, if one's goal is to grow relatively small, straight stems for use in construction or as fuel wood or if one is planting in areas where waves and other forms of environmental disturbance exact a significant toll on survival (Yao 1996; Pedro Balagas and Emma Melana, personal communication). The dominant and most highly valued use of planted mangroves in Bais is for posts used in the construction of traditional fish corrals, called *bunsod*. For this purpose, tree stems must be straight and are typically harvested when small, i.e., between 2.5 and 5.0cm dbh (Walters 2000a). If spacing is too close, trees will become crowded early and grow slowly. If spacing is too wide, trees will manifest bushy growth and are more likely to have crooked stems.

13 In fact, there is a highly significant positive correlation between plantation size and spacing used (R=0.478, F=26.02, p<0.001, df=89).

14 Newly planted mangroves can be killed by barnacles and oyster spat which attach and grow on stems, often in such densities that young stems are either physically toppled by the weight of the growing shells or are asphyxiated by having photosynthetic surfaces on their stems blocked by the growing layer of shells (Cabahug et al. 1986; DENR 1994). Vulnerability to damage from shell infestations decreases rapidly as the stem strengthens with age and grows leaves. Planted trees are typically most vulnerable in their first six months.

15 Sediment deposition near the mouths of rivers in Bais is often substantial and results from upland soil erosion (Calumpong & Luchavez 1997). Near the mouth of the Tamugong River in Bais Bay, for example, sedimentation during the past four decades has raised the topographic level of offshore lands sufficiently to expand the habitat suitable for mangroves by nearly 200 ha (Walters 2000a).

16 Persons fishing, gleaning, or passing their boats along the shore often destroy young trees. In most cases, such damage is unintended, but some deliberate destruction of planted seedlings in protest to having common areas planted was also found to have occurred (Walters 1998).

17 In addition to expanding populations of fisherfolk, landless families from surrounding areas now live along the shoreline around Bais Bay. Former village boundaries have been blurred in many areas as a result of increase in the density of houses.

18 See Vayda 1995: 225–229 for further discussion of how design analysis and comparative studies may sometimes be used, in the absence of direct historical evidence, to distinguish those beneficial consequences that are by-products or the effects of chance from those that occur because particular behavioural traits have been fashioned by some selective process, such as trial-and-error, to produce them.

19 For a discussion and illustrations of how our having substantial antecedent knowledge of a local people's culture is sometimes but not always helpful when we are looking for causes of events in which the people are involved, see Vayda & Sahur 1996:51.

20 It may be noted incidentally that our arguments here are in opposition also to Sahlins's decades-long insistence that the cultural must be the object of the 'anthropological project' (see, for example, Sahlins 1976, 1977, 1995).

References

Alcorn, J. 1989. Process as resource: The traditional agricultural ideology of Bora and Huastec resource management and its implications for research. In *Resource management in Amazonia: Indigenous and folk strategies. Advances in economic botany.* 7 (eds.) D.A. Posey & W. Balee. New York: The New York Botanical Garden. 63–77.

Baconguis, S.R., D.M. Cabahug & S.N. Alonzo-Pasicolon. 1990. Identification and inventory of Philippine forested-wetland resource. *Forest Ecology and Management.* 33/34. 21–44.

Barth, F. 1987. *Cosmologies in the making: A generative approach to cultural variation in inner New Guinea.* Cambridge: Cambridge University Press.

Barth, F. 1994. A personal view of present tasks and priorities in cultural and social anthropology. In *Assessing cultural anthropology.* (ed.) R. Borofsky. New York: McGraw-Hill. 349–361.

Becht, C.J.G. 1939 Rattenplaag-Bezwering in Goenoengkidoel. *Djawa* 19: 258–259.

Bentley, J. W., & K. L. Andrews 1991. Pests, peasants, and publications: Anthropological and entomological views of an integrated pest management program for small-scale Honduran farmers. *Human Organization* 50: 113–124.

Berkes, F. 1999. *Sacred ecology: Traditional ecological knowledge and resource management.* Philadelphia: Taylor & Francis.

Bratman, M. E. 1992. Practical reasoning and acceptance in a context. *Mind* 101: 1–15.

Brown, W.H. & A.F. Fischer. 1918. *Philippine mangrove swamps.* Bulletin No. 17. Manila: Bureau of Forestry, Department of Agriculture and Natural Resources.

Cabahug, D.M. Jr., F.M. Ambi, S.O. Nisperos & N.C. Truzan Jr. 1986. Impact of community-based mangrove forestation to mangrove dependent families and to nearby coastal areas in Central Visayas: A case example.' In *Mangroves of Asia and the Pacific: Status and management.* (ed.) National Mangrove Committee. Quezon City, Philippines: Natural Resources Management Center, Ministry of Natural Resources. 441–466.

Calumpong, H.P. 1994. Status of mangrove resources in the Philippines. In *Proceedings of the Third ASEAN-Australia Symposium on Living Coastal Resources: Volume 1.* (eds.) C. Wilkinson, S. Sudara & C.L. Ming. Townsville: Australian Agency for International Development (AUSAID) and Australian Institute of Marine Science. 215–229.

Calumpong, H.P., & J.A. Luchavez. 1997. Profile of the Bais Bay Basin. *Silliman Journal.* 37 (3/4): 1–27.

Carlyle, T. 1899. *Critical and Miscellaneous Essays.* Vol. 2. London: Chapman & Hall.

Chambers, R. 1991. Shortcut and participatory methods for gathering social information for projects. In *Putting people first: Sociological variables in rural development.* (2nd ed). (ed.) M. Cernea. New York: Oxford University Press. 515–537.

Cosmides, L. & J. Tooby. 1996. Are humans good intuitive statisticians after all? Rethinking some conclusions from the literature on judgment under uncertainty. *Cognition* 58: 1–73.

Department of Environment and Natural Resources (DENR) 1990. *Compilation of mangrove regulations.* Quezon City, Philippines: Coastal Resources Management Committee, Republic of the Philippines.

Department of Environment and Natural Resources (DENR) 1994. *Mangrove regeneration and management.* Mangrove Technical Review Committee, Fisheries Sector Program. Quezon City: Republic of the Philippines (DENR).

Dewalt, B.R., P. Vergne & M. Hardin. 1996. Shrimp aquaculture development and the environment: People, mangroves and fisheries on the Gulf of Fonseca. Honduras. *World Development* 24: 1193–1208.

Diop, E.S. (ed.) 1993. *Conservation and sustainable utilization of mangrove forests in Latin America and African regions (Part 2: Africa). Mangrove Ecosystem Technical Reports 3.* Okinawa: International Society for Mangrove Ecosystems and International Tropical Timber Organization.

Ellen, R. 1998. Comment [on Sillitoe 1998a]. *Current Anthropology* 39: 238–239.

Food and Agriculture Organization of the United Nations (FAO) 1994. *Mangrove forest management guidelines. FAO Forestry Paper 117.* Rome: Food & Agriculture Organization of the United Nations.

Gallagher, K. n.d. *Training model of the farmers' field guide.* Jakarta and Yogyakarta: Indonesian National IPM Programme.

Gallagher, K.D., P.E. Kenmore & K. Sogawa. 1994. Judicial use of insecticides deters planthopper outbreaks and extends the life of resistant varieties in Southeast Asian rice. In *Planthoppers: Their ecology and management.* (eds.) R. Denno & T.J. Perfect. London: Chapman & Hall. 599–614.

Gatewood, J.B. 1985. Actions speak louder than words. In *Directions in cognitive anthropology.* (ed.) J.W.D. Dougherty. Urbana: University of Illinois Press. 199–219.

Gigerenzer, G. 1991. How to make cognitive illusions disappear: Beyond 'heuristics and biases.' *European Review of Social Psychology.* 2: 83–115.

Gigerenzer, G. 1992. *Cognitive illusions illusory? Rethinking judgment under uncertainty.* Preprint Series of the Research Group on Biological Foundations of Human Culture (No.

25/92, 1991/92), Bielefeld, Germany: Center for Interdisciplinary Research, University of Bielefeld.

Goodenough, W.H. 1957. Cultural anthropology and linguistics. In Report of the seventh annual round table meeting on linguistics and language study. (ed.) P.L. Garvin. Washington: Georgetown University Press. 167–177.

Goodenough, W.H. 1994. Toward a working theory of culture. In *Assessing cultural anthropology*. (ed.) R. Borofsky. New York: McGraw-Hill. 262–275.

Grist, D.H. 1986. *Rice* (6th ed). London: Longman.

Hamilton, L.S., J.A. Dixon & G.O. Miller. 1989. Mangrove forests: An undervalued resource of the land and of the sea. In *Ocean yearbook 8*. (eds.) E.M. Borgese, N. Ginsburg & J.R. Morgan. Chicago: University of Chicago Press. 254–288.

Harris, M. 1968. *The rise of anthropological theory*. Crowell: New York.

Hawthorn, G. 1991. *Plausible worlds: Possibility and understanding in history and the social sciences*. Cambridge: Cambridge University Press.

Hess, C. 1997. *Hungry for hope: On the cultural and communicative dimensions of development in highland Ecuador*. London: Intermediate Technology Publications.

Hunn, E. 1977. *Tzeltal folk zoology*. New York: Academic Press.

Hunn, E. 1989. Ethnoecology: The relevance of cognitive anthropology for human ecology. In *The relevance of culture*. (ed.) M. Freilich. New York: Bergin & Garvey. 143–160.

Hunn, E. 1993. What is traditional ecological knowledge? In *Traditional ecological knowledge: Wisdom for sustainable development*. (eds.) N.M. Williams & G. Baines. Canberra: Centre for Resource and Environmental Studies, Australian National University. 16–20.

Huntington, H.P. 2000. Using traditional ecological knowledge in science: Methods and applications. *Ecological Applications* 10: 1270–1274.

Hutchins, E. 1995. *Cognition in the wild*. Cambridge, MA: MIT Press.

Johnson, A. 1974. Ethnoecology and planting practices in a swidden agricultural system. *American Ethnologist* 1:209–223.

Kaly, U.L., & G.P. Jones. 1998. Mangrove restoration: A potential tool for coastal management in tropical developing countries. *Ambio* 27: 656–661.

Keller, C., & J.D. Keller. 1993. Thinking and acting with iron. In *Understanding practice: Perspectives on activity and context*. (eds.) S. Chaiklin & J. Lave. Cambridge: Cambridge University Press. 125–143.

Kenmore, P.E. 1991. *Indonesia's integrated pest management – A model for Asia*. Manila (FAO P.O. Box 1864): FAO Rice IPC Programme.

Kenmore, P.E., J.A. Litsinger, J.P. Bandong, A.C. Santiago & M.M. Salac. 1987. Philippine rice farmers and insecticides: Thirty years of growing dependency and new options for change. In *Management of pests and pesticides: Farmers' perceptions and practices*. (eds.) J. Tait & B. Napompeth. Boulder: Westview Press. 98–108.

Khoon, G.W., & O.J. Eong. 1995. The use of demographic studies in mangrove silviculture. *Hydrobiologia* 295: 255–261.

Kunstadter, P., E.C.F. Bird & S. Sabhasri. (eds.) 1986. *Man in the mangroves*. Tokyo: United Nations University.

Lacerda, L.D. (ed.) 1993. Conservation and Sustainable Utilization of Mangrove Forests in Latin America and African Regions (Part 1: Latin America*). Mangrove Ecosystem Technical Reports 2*. Okinawa: International Society for Mangrove Ecosystems and International Tropical Timber Organization.

Last, M. 1981. The importance of knowing about not knowing. *Social Science and Medicine* 15B: 387–392.

Lave, J. 1988. *Cognition in practice*. Cambridge: Cambridge University Press.

Lees, S. 2000. Review of Hess 1997. *American Ethnologist* 27: 503–504.

Lewis, D. 1986. *Philosophical Papers, Vol. 2*. New York: Oxford University Press.

Lewis, R.R. 1990. Creation and restoration of coastal plain wetlands in Florida. In *Wetland creation and restoration: the status of the science*. (eds.) J.A. Kusler & M.E. Kentula. Washington: Island Press. 73–101.

Macnae, W. 1968. A general account of the fauna and flora of mangrove swamps and forests in the Indo-West-Pacific region. *Advances in Marine Biology* 6:73–270.

Peirce, C.S. 1932. *Collected Papers of Charles Sanders Peirce*, Vol. 2. (eds.) C. Hartshorne & P. Weiss. Cambridge: Harvard University Press.

Pelkey, N. 1995. Please stop the PRA RRA rah. *Out of the Shell* (Coastal Resources Research Network Newsletter) 5 (1): 17–24.

Pomeroy, R.S., R.B. Pollnac, C.D. Predo & B.M. Katon. 1996. *Impact evaluation of community-based coastal resource management projects in the Philippines*. Makati City, Philippines: International Center for Living Aquatic Resources Management (ICLARM).

Portes, A., & J. Sensenbrenner. 1993. Embeddedness and immigration: Notes on the social determinants of economic action. *American Journal of Sociology* 98: 1320–1350.

Primavera, J.H. 1995. Mangroves and brackish water pond culture in the Philippines. *Hydrobiologia* 295: 303–309.

Primavera, J.H., & R.F. Agbayani. 1996. Comparative strategies in community-based mangrove rehabilitation programs in the Philippines. Paper presented at the ECOTONE V conference: Community Participation in Conservation, Sustainable Use and Rehabilitation of Mangroves in Southeast Asia. Ho Chi Minh City, Vietnam, January 8–12.

Purcell, T.W. 1998. Indigenous knowledge and applied anthropology: Questions of definition and direction. *Human Organization* 57: 258–272

Quinn, N., & D. Holland. 1987. Culture and cognition. In *Cultural models in language and thought*. (eds.) D. Holland & N. Quinn. New York: Cambridge University Press. 3–40.

Rheingold, H. 1988. *They have a word for it*. Los Angeles: J.P. Tarcher.

Rorty, R. 1989. *Contingency, irony, and solidarity*. Cambridge: Cambridge University Press.

Saenger, P., & N.A. Siddiqi. 1993. Land from the sea: The mangrove afforestation program in Bangladesh. *Ocean and Coastal Management* 20: 23–39.

Sahlins, M. 1976. *Culture and practical reason*. Chicago: University of Chicago Press.

Sahlins, M. 1977. The state of the art in social/cultural anthropology: The search for an object. In *Perspectives on anthropology* 1976. (eds.) A.F.C. Wallace, J.L. Angel, R. Fox, S. McLendon, R. Sady & R. Sharer. Special Publication No. 10. Washington: American Anthropological Association.

Sahlins, M. 1995. *How 'natives' think: About Captain Cook, for example*. Chicago: University of Chicago Press.

Settle, W. H., H. Ariawan, E.T. Astuti, W. Cahyana, A.L. Hakim, D. Hindayana, A.S. Lestari & Pajarningsih. 1996. Managing tropical rice pests through conservation of generalist natural enemies and alternative prey. *Ecology* 77: 1975–1988.

Shepherd, K.R. 1986. *Plantation silviculture*. Boston: Martinus Nijhoff Publishers (Kluwer Academic Publishers).

Siddiqi, N.A., & M.A.S. Khan. 1990. Growth performance of mangrove trees along the coastal belt of Bangladesh. *Mangrove Ecosystems Occasional Papers* No. 8(1), UNESCO/UNDP/COMAR.

Sillitoe, P. 1996a. *A place against time: Land and environment in the Papua New Guinea*. Amsterdam : Harwood Academic.

Sillitoe, P. 1998a. The development of indigenous knowledge: A new applied anthropology. *Current Anthropology* 39: 223–252.

Sillitoe, P. 1998b. It's all in the mound: Fertility management under stationary shifting cultivation in the Papua New Guinea highlands. *Mountain Research and Development* 18: 123–134.

Sillitoe, P. 1998c. What, know natives? Local knowledge in development. *Social Anthropology* 6: 203–220.

Sukardjo, S., & I. Yamada 1992. Biomass and productivity of a *Rhizophora mucronata* Lamarck plantation in Tritih, Central Java, Indonesia. *Forest Ecology and Management* 49: 195–209.

Sunley, P. 1996. Context in economic geography: The relevance of pragmatism. *Progress in Human Geography* 20: 338–355.

Thagard, P. & C. Shelley. 1997. Abductive reasoning: Logic, visual thinking, and coherence. In: *Logic and Scientific Methods.* (eds.) M.L. Dalla Chiara, K. Doets, D. Mundici, & J. van Benthem. Dordrecht: Kluwer. 413–427.

Thomason, R. H. 1987. The context-sensitivity of belief and desire. In *Reasoning about actions & plans.* (eds.) M.P. Georgeff & A.L. Lansky. Los Altos: Morgan Kaufmann. 341–360.

Thorhaug, A. 1990. Restoration of mangroves and seagrasses – economic benefits for fisheries and mariculture. In Environmental restoration: Science and strategies for restoring the earth. (ed.) J.J. Berger. Washington: Island Press. 265–281.

Tomlinson, P.B. 1986. *The botany of mangroves.* London: Cambridge University Press.

Van de Fliert, E. 1993. *Integrated pest management: Farmer field schools generate sustainable practices. A case study in Central Java evaluating IPM Training.* Wageningen Agricultural University Papers No. 93-3. Wageningen, The Netherlands: Wageningen Agricultural University.

Van Speybroeck, D. 1992. Regeneration strategy of mangroves along the Kenya coast: A first approach. *Hydrobiologia* 247: 243–251.

Vayda, A.P. 1983. Progressive contextualization: Methods for research in human ecology. *Human Ecology* 11: 265–281.

Vayda, A.P. 1994. Actions, variations, and change: The emerging anti-essentialist view in anthropology. In *Assessing cultural anthropology.* (ed.) R. Borofsky. New York: McGraw-Hill. 320–330.

Vayda, A.P. 1995. Failures of explanation in Darwinian ecological anthropology: Part I. *Philosophy of the Social Sciences* 25: 219–249.

Vayda, A.P. 1996. *Methods and explanations in the study of human actions and their environmental effects.* CIFOR/WWF Special Publication. Jakarta: Center for International Forestry Research (CIFOR).

Vayda, A.P. 1997. *Managing forests and improving the livelihoods of forest-dependent people: Reflections on CIFOR's social science research in relation to its mandate for generalisable strategic research.* CIFOR Working Paper No. 16. Jakarta: Center for International Forestry Research (CIFOR).

Vayda, A.P. 1998. Anthropological perspectives on tropical deforestation? A review article. *Anthropos* 93: 573–579.

Vayda, A.P., & A. Sahur 1996. *Bugis settlers in East Kalimantan's Kutai National Park: Their past and present and some possibilities for their future.* A CIFOR Special Publication. Jakarta: Center for International Forestry Research (CIFOR).

Vayda, A.P., & I. Setyawati 1998. Questions about culture-related considerations in research on cognition and agro-ecological change: Illustrations from studies of agricultural pest management in Java. *Antropologi Indonesia* (Indonesian Journal of Social and Cultural Anthropology), No. 55: 44–52.

Vayda, A.P., & B.B. Walters. 1999. Against political ecology. *Human Ecology* 27: 167–179.

Von Wright, G.H. 1971. *Explanation and Understanding.* Ithaca: Cornell University Press.

Waddell, E. 1972. *The mound builders: Agricultural practices, environment, and society in the Central Highlands of New Guinea.* Seattle: University of Washington Press.

Waddell, E. 1973. Raiapu Enga adaptive strategies: Structure and general implications. In *The Pacific in transition: Geographical perspectives on adaptation and change.* (ed.) H.C. Brookfield, New York: St. Martin's Press. 25–54.

Waddell, E. How the Enga cope with frost: Responses to climatic perturbations in the Central Highlands of New Guinea. *Human Ecology* 3: 249–273.

Wadsworth, F.H. 1997. *Forest production for tropical America. Agricultural handbook 710.* Washington: U.S. Department of Agriculture (Forest Service).

Walters, B.B. 1995. People, policies and resources: Mangrove restoration and conservation in the Bais Bay Basin, Negros Oriental and wider Philippine context. In *Philippine coastal resources under stress.* (eds.) M.A. Juinio-Menez and G.F. Newkirk. Halifax, Canada: Coastal Resources Research Network and Quezon City, Philippines: Marine Science Institute. 151–165.

Walters, B.B. 1997. Human ecological questions for tropical restoration: Experiences from planting native upland forest and coastal mangrove trees in the Philippines. *Forest Ecology and Management* 99: 275–290.

Walters, B.B. 1998. Muddy mangroves and murky common property theories. Paper presented at the 1998 meeting of the International Association for the Study of Common Property Resources, Vancouver, Canada, June 6–9.

Walters, B.B. 2000a. Event ecology in the Philippines: Explaining mangrove cutting and planting and their environmental effects. Ph.D. dissertation, Rutgers University, New Brunswick, New Jersey.

Walters, B.B. 2000b. Local mangrove planting in the Philippines: Are fisherfolk and fish pond owners effective restorationists? *Restoration Ecology* 8 (3): 237–246.

Walters, B.B., A.M. Cadelina, A. Cardano & E. Visitacion. 1999. Community history and rural development: why some farmers participate more readily than others. *Agricultural Systems* 59: 193–214.

Weinstock, J.A. 1994. *Rhizophora* mangrove agroforestry. *Economic Botany* 48: 210–213.

Yao, C.E. 1996. Mangrove reforestation in the Central Visayas. *Greenfields* 24 (5/6): 29–31, 34–37.

Yao, C.E., & F. Nanagas 1984 Banacon Island: Biggest bakauan plantations in Central Visayas. Canopy International 10 (July-Sept.): 1, 4, 5, 9.

Chapter 3

A Decision Model for the Incorporation of Indigenous Knowledge into Development Projects

Paul Sillitoe and Julian Barr

Agencies increasingly accept that the incorporation of indigenous knowledge into programmes and projects will advance development agendas. As Warren noted over a decade ago, "Recent publications by The World Bank reflect the appreciation of the cost-effective role indigenous knowledge systems can play in the development process" (1991: 29). The question they are more and more asking is how they can achieve this effectively (and one should add, ethically, although we do not discuss this here – see Laird 2002 for ethical discussion in context of biodiversity conservation). The decision model advanced here[1] shows how agencies can make connections between local people's understandings and practices, and research and development approaches in a range of sectors – education, environment, health and population, social development, infrastructure and urban planning, and natural resources. It suggests how to go about:

- facilitating meaningful communication with local people;
- understanding and valuing local people's knowledge and practices;
- identifying research or development initiatives to address jointly perceived problems;
- advancing participatory research to identify and tackle constraints more effectively; and
- discussing with people the social, environmental and other consequences of any action.

The decision model illustrates various strategies for collaborative, interdisciplinary research, and methods to further incorporate indigenous knowledge into development initiatives. It supplements and advances on methodologies formulated by others (e.g. Emery 2000, Grenier 1998; Gustafsson 1995, IRRI 1996, Purcell & Onjoro 2002, Rajasekaran 1994). It assumes that managers accept that by paying attention to local perceptions and practices, research and development initiatives are more likely to be relevant to people's needs and generate sustainable interventions, local stakeholders drawing on their knowledge in partnership with researchers to create new understandings designed to have practical and beneficial outcomes for the poor.

This chapter focuses on the practical and technical issues of designing and managing projects that incorporate an indigenous knowledge component, leaving aside intellectual debates on the meaning of indigenous knowledge, the desirability, even propriety of such approaches to development, and so on (see Sillitoe 1998, 2000, 2002b). The aim is to explore how to achieve an appropriate balance. It is not prescriptive, but provides a range of options from which a project manager can choose the most suitable to meet the demands of any particular project. It assumes a project manager who is convinced of the merit of incorporating indigenous knowledge into a project but is uncertain of how to go about it given project goals and resource limitations, and requires some guide to make informed choices about the techniques and options that are available. The chapter seeks to lead such persons through the key issues in the trade-off between the pragmatic and the ideal in project design and management by providing some aids to project design, accompanied by some debate on the consequences of making particular choices. We take the principal design issues to concern resources, time, and project scope. We conceive of these as axes on what we call a project design cube, a new concept that we think will assist in the management of development projects that seek to incorporate indigenous knowledge. The three design issues impact on staffing, and data collection and analysis that a project can undertake. There is a problem of synchronisation in such interdisciplinary team projects, particularly those that attempt to combine social and natural sciences. It is preferable to commence a cycle of indigenous knowledge study in advance of other work, the results of this indigenous knowledge research feeding into the planning phase of related projects. We call this arrangement the project wave, another new concept intended to assist in the management of development projects that seek to incorporate indigenous knowledge.

Some methodological issues

Indigenous knowledge work is not straightforward and it is well to have some awareness of the considerable problems that attend this work (Sillitoe 2002a: 15). It is necessary to proceed cautiously. Managers of projects with an indigenous knowledge component should employ staff aware of difficulties and methods devised to tackle them. These demand further attention to integrate indigenous knowledge into the development process and the model advanced here makes no pretence at advancing on them. It works within current constraints to formulate strategies that can help meet the demands of development now – cost-effective, time-effective, generating relevant insights, readily intelligible to non-experts etc. – while not downplaying the difficulties so as to render the work effectively valueless. While managers will assess attempts to advance on current techniques according to their resource effectiveness, they should set these demands against the range of data collected and its reliability.

A key problem is facilitating meaningful *communication* between development personnel and local people to establish what each has to offer, informing science with ethnographic findings about people's knowledge and locals about the scope of science and what it might offer, so that they can better understand the alternatives available in addressing problems, so realising the comparative advantages of each. The promotion of more effective participation in the identification and researching of problems can only be achieved so far as awareness, knowledge and socio-political barriers will allow. We seek a methodology that allows both outsiders and insiders to contribute as necessary, balancing between technocrats defining the problem/constraint, which is arrogant and ethnocentric, and the local people doing so, which hits cultural barriers that thwart scientific research.

It is necessary to promote a *collaborative* atmosphere in which neither scientific nor local interests feel threatened, assuring all parties that they have a role in negotiations, with vital skills and knowledge. This implies demonstrating how awareness of indigenous knowledge will improve the relevance of development work and vice versa. The absence of a coherent indigenous knowledge approach that might interface effectively with science and technology is a limitation, contributing to scientists failing to appreciate how it might inform their research. The presentation of indigenous knowledge in a manner accessible to others, such that they can see its relevance to their work, means avoiding jargon-loaded and obscure accounts, while not overlooking insights gained in cross-cultural research, often in subtle arguments. There is a need to avoid oversimplification of complex issues, inviting distortion and misrepresentation in the search for user-friendly accounts.

The advancement of *interdisciplinary* work is central to indigenous knowledge research, particularly combining the technical know-how of natural scientists with the empathy of social scientists (Cochrane 1971; Sillitoe *et al.* 2002). An integrated perspective implies a willingness to learn from one another, in addition to local people. The indigenous knowledge component of any research and/or development project should not necessarily dominate. There must be a genuine two-way flow of ideas and information between all parties. Motivation depends in considerable measure on fostering consensus decisions, joint ownership and open debate. Indigenous knowledge research should maintain a wide socio-cultural perspective to contextualize the narrowly focused work of technical specialists. In science-speak, we cannot understand cultures by looking at individual parts in isolation, as complex systems they manifest emergent properties that we can only see when all the parts are working together. It is not possible to predict which cultural domains might relate intimately with others, often unexpected practices impinge on one another.

The *dynamism* of indigenous knowledge presents further difficulties (Warren 1991). As indigenous knowledge is neither static nor uniform, it cannot be documented once-and-for-all, but is subject to continual negotiation between stakeholders. If we hope to accommodate to the dynamic nature of local knowledge we need an iterative research strategy, closely linking development interventions to on-going indigenous knowledge investigations. After all, development aims to accelerate change, dramatically modifying indigenous knowledge with scientific perspectives. We need beware of indigenous knowledge research contributing to the accumulation of exotic ethnographic data that are sterile and undynamic from a developmental perspective, even potentially disempowering people by representing their knowledge in ways inaccessible to them and beyond their control, maybe infringing their intellectual property rights (Brush & Stabinsky 1996).

The *one-offness* of indigenous knowledge hampers its deployment in development, impeding the formulation of generalisations that might inform wider policy and practice. Its small-scale, culturally specific and geographically local nature hinders the advancement of an integrated approach. (A possible red herring given the variety of knowledge traditions worldwide, their internal variations regarding individuals' unsystematised understandings and their constant revision over time.) This variation makes generalisation potentially dangerous, imputing ideas elsewhere that may be inappropriate. Nonetheless we need to evolve principles that will facilitate a degree of reliable generalisation from indigenous knowledge research, to go beyond local case studies (often small NGOs working in a few communities with appropriate technology interventions), which are not cost effective to replicate in large numbers.

At first sight indigenous knowledge work seems straightforward enough, we just have to ask some local culture bearers what they think. But we soon run into *cross-cultural problems* that challenge what we think we know. Development oriented indigenous knowledge work is no different to any other ethnographic enquiry in this respect. The task of sympathetically accessing concepts in local usage, and conveying something about them, is large. Knowledge is diffuse and communicated piecemeal in everyday life. Local stakeholder knowledge is not homogenous. There is often no consensus among the 'natives'. People transfer much knowledge through practical experience, and are unfamiliar with expressing all that they know in words. They may also carry knowledge, and transfer it between generations, using alien idioms featuring symbols, myths, rites and so on. Translating what we hear into foreign words and concepts further misconstrues whatever it is that we manage to comprehend about other's views and actions. Understanding is inevitably limited given our outsider perspective.

The *time scale* required in ethnographic research is normally considerable which presents problems in development contexts with their short-term politically driven considerations demanding quick returns. Managers need to understand that indigenous knowledge research is usually long-term. It can take several years, not months or weeks, for someone unacquainted with a region to achieve meaningful insight into local knowledge and practices, and from this perspective inform development projects. The understanding that can be accomplished in a single project cycle will be of a different order. While some indigenous knowledge research may be attempted in short time frames, as our decision model allows, it is necessary to be aware of the costs of necessary compromises. It is not just a question of the time it takes to learn language, cultural repertoire, social scenario and so on, but also the investment needed to win the trust and confidence of people who frequently have reason to be extremely suspicious of foreigners and their intentions.

The indigenous knowledge project decision cube

The design of projects that incorporate indigenous knowledge is often a balancing act between what is intellectually advisable and what is feasible with available resources. The decision model provides non-prescriptive guidance on achieving an appropriate balance in designing a project. It takes managers through the key questions and trade-offs and points out the consequences of decisions. It seeks to inform choices between available options and techniques under specified resource limitations. The principal resource issues are *cost, time* and *scope* of objectives, which we present as axes on a project design cube. These impact on staffing, as the main input in indigenous knowledge projects, which informs range of data collected and analysis according to staff qualifications. The cube device may be used to inform decision making:

- From the start of a project, to design an integrated indigenous knowledge contribution to a project, setting budget, timeframe and objectives to achieve required results.
- During a project, when realised that an indigenous knowledge component would help; budget and time frame set, go to the appropriate cube and see what staffing, research and results are achievable.

Managers planning to incorporate an indigenous knowledge component in a project must address three questions before they can make decisions about staffing, methods, outputs etc.:

- What *resources* are available – relates to staff?
- How much *time* is available – relates to project cycle management?
- What is the *scope* of the indigenous knowledge inquiry – relates to outputs?

These are continuous variables and we can represent each as one of the three dimensions of a cube, which gives us our indigenous knowledge decision cube.

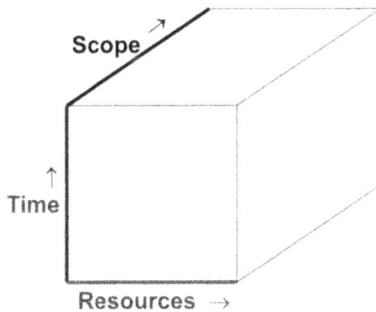

Figure 3.1   Indigenous knowledge project design variables as cube axes

The resources axis

The first task is to specify financial resources on the budget axis, as the sum available to finance the indigenous knowledge work will determine research undertaken. The major cost will be staff. Indigenous knowledge research has relatively low equipment costs (laptops, cameras, tape recorders etc.), although these increase with demands for results other than written data such as digital audio and video records, and sophisticated computer data analyses demanding transcription services. Other expenditure will be on travel and subsistence costs.

Most projects will calculate to employ a combination of staff; say 10% of a lecturer's time, a full-time post-graduate student plus three research assistants. University staff may make a full-time commitment, and commonly expect to supervise a team, generally comprising students. They may leave post-doctoral researchers to work on their own for considerable periods. Consultants are likely to work singly. Managers also need to balance expenditure against project scope. If the aim is narrow – e.g. collecting indigenous knowledge on limited aspects of technology – funding a 'major' project is unlikely to prove value for money.

The quality of indigenous knowledge research will increase in relation to the project budget, increasing the time staff can interact with local people. The short contact periods achievable on poorly resourced projects will result in shallow descriptive indigenous knowledge data. A certain amount of time needs to be spent interacting with local people to achieve a minimally acceptable understanding of local issues, beyond this threshold understanding increases. Better resourced projects can also afford more senior staff, with the experience to push understanding of indigenous knowledge deeper. There is no substitute for 'qualified time' in achieving reliable outputs. For example, the data collected by ten research assistants is unlikely to be ten times better than that by one research assistant, whereas this type of relationship may well pertain when purchasing additional senior staff time.

Staff issues

A range of staff may undertake indigenous knowledge research, ranging in experience from novices – such as undergraduates and young field staff – to experts – such as senior NGO staff, consultants and university professors (see list Table 3.1). Any project personnel with some familiarity with participatory methods should be able to contribute to indigenous knowledge research whatever their disciplinary background. Staff will be found largely in the university/research institute sector, either in-country or overseas. The consultancy and NGO sectors may also supply suitable persons. Different staff are appropriate to different types of indigenous knowledge research; for example academics and consultants have contrasting advantages and disadvantages – see Table 3.3.

Factors informing staff selection include availability, qualifications, familiarity with indigenous knowledge methods, previous experience in locality, language ability and cost of employment, and will vary with project aims. Table 3.1 gives strengths, weaknesses, opportunities and constraints of different staff, for managers to appraise options according to the three indigenous knowledge-project-decision-cube axes: cost of employing personnel (resources), availability (time), and research expertise (scope). Managers will seek to balance cost against project aims, matching staff expertise to quality of data sought. While available resources set limits, they have room for manoeuvre.

Managers should also note the following

Many problems are attributable to time constraints, not teamwork failures. What competition is there for individuals' time? Academics are likely to have other commitments (research, teaching, administration, supervision etc.) and consultants too (other projects – though multiple involvement can add value, informing methods and ideas, over-commitment can result in chronic under performance). How good are individuals' time management skills to handle such competing demands? There is a danger of trust breaking down if staff fail to meet fully their obligations to a project due to over-commitment, others become reluctant to commit themselves to work that requires the co-operation of persons on whom they think they cannot rely. Short term contracts may inhibit staff from engaging with indigenous knowledge components of projects, compelling them to focus on completing standard contracted inputs quickly and moving on.

What are the incentives to perform: financial rewards, travel opportunities, equipment budget, professional recognition, chance to further research, opportunities to publish, conferences attendance? Managers should ensure that the project can deliver expected rewards.

An indigenous knowledge research team of mixed background comprising both nationals and foreigners will probably work best, staff more likely to cover for weaknesses in one another. Projects need to work at fostering an interdisciplinary research environment.

Local staff may more readily access social networks; they may be part of them. But they will be more constrained by local institutions, social expectations and politics than outsiders, their work skewed as participants (Ahmed 2000). They may further empowerment objectives, encouraging local people to express and collate indigenous knowledge themselves. National social scientists may also rapidly enter local networks. Local ideas may entirely inform understanding of technical issues – such staff are likely to be genuinely naïve in respect of international science. Their studies run the risk of becoming 'too ethnographic' and failing to focus on the technical dimensions of development research. But partnered with someone with a technical background and appreciation of anthropological methods such staff should prove excellent value in indigenous knowledge enquiries. Once focussed on the technical problems, their data should prove robust, explore issues in depth, and address socio-cultural dimensions, and probe local understanding of problems. Such staff may face personal problems interacting with other project personnel who may perceive them and their work as low status (Alam 2000). National technicians/ scientists are likely to have problems with open-ended ethnographic methods being familiar with more controlled investigations. They will tend to transform indigenous knowledge into scientific contexts.

Foreign scientists and technicians may exacerbate such problems, even those sympathetic to indigenous knowledge enquiries. They are likely to be chary of ethnographic methods, for failing to yield 'real' (i.e. quantitative) data. – a manifestation of the tension between natural and social sciences. They are likely to press on with scientific research; impatient at the time it takes to deliver indigenous knowledge relevant to their work. They may contractually be in no position to entertain experimenting with non-standard research methods. They are unlikely to engage meaningfully with the local culture. Most projects interested in so engaging and accessing indigenous knowledge will employ a social scientist, often an anthropologist. The ideal is someone with an interdisciplinary background who can understand technical issues and converse with scientists on their own terms, and so promote acceptance of indigenous knowledge (den Biggelaar 1991). Some staff options (e.g. such specialists, senior academics etc.) may not be available even if the budget is sufficient. Students may prove good value for indigenous knowledge work. While tenured academics, short-term consultants and local field assistants may only undertake work while remunerated, students, investing in their futures with qualifications, are likely to work diligently throughout and beyond a project.

The time axis

The decision cube time axis is most likely to extend between 3 months and 3 years, any indigenous knowledge study having to fit in the project cycle (ODA 1988:3–12; Nolan 2002:95–103). The time budgeted will determine the type of project (Table 3.4). Time and project scope are linked, as the longer researchers can spend gathering and analysing data the higher the quality of outputs.

Short time frames have implications for quality of research:

- Shallow interaction with people.
- Limit data collection methods to the more rapid options.
- Limit options for data analysis.
- Encourage short-term employment contracts.
- Inhibit meaningful commitment to project.
- Lack of staff continuity.
- Chronic job insecurity limits project team building.

Indigenous knowledge studies of very limited scope (e.g. focussing on a single technological issue) may be completed in short time frames. Many problems relate to time, agencies demanding too much of researchers in unrealistic time spans.

Table 3.1 Qualifications and suitability analysis of different staff

Code	Staff type	Status	Experience/ qualifs.	Methods (can use)	Output quality/ relevance	Availability Academic	Availability Consultant	Cost	Strengths	Weaknesses	Opportunities	Constraints
1	village enumerators – trained in IK work	J/M	- to + 1y/2y school	S	V		F/P	a	Immersed in local culture. Speak language	Inexperienced Lacks outsider's objective view.	Time is readily available Focus on output tasks	Bound by cultural mores/ biases
2	local ethnographer	J/M	+ to +++	S-M	V – MH		F/P	a	May know key informants	Little / no formal education	Provide long-term contact	
3	junior national NGO workers with social science training	J/M	+ ~BA	S-M	V – A		F/P	a				
4	senior national NGO workers with social science training	M/S	++ to +++ ~BA	S-D	A – MH		F/P	a/b-c	Basic sociology training	Variable experience Variable project commitment	Time for focused work May have regular contact with community	Influenced by cultural mores/ biases
5	national undergraduate sociologist	J	-	S	V	PT-1, >1m	-	a				
6	national graduate sociologist	J	+ BA	S-M	V – A		FT	a				
7	foreign undergraduate anthropologist	J	-	S	V	PT-1†, >1m	-	a	Basic anthro-pology training	Inex-perienced No influence, difficulties with respondents	Keen Time for focused study	Culture shock Language Illness
8	foreign graduate anthropologist[1]	J/M	+ BA	S-M	V-A	FT	FT	a/b				

9	national graduate scientist with interest in IK work	J/M	+ BSc	S-M	V – A		FT	a	Technologically knowledgeable Know society & language	Experience and commitment to project variable	Time for research Ready access to local culture	Disdain for local customs Possibly arrogant
10	national postgraduate scientist with interest in IK work	J/M	+ to ++ MSc/PhD	S-D	A – MH	F/P-3/4, >1m	FT	a		Inexperienced No influence Naïve regarding IK approach	Keen Time for detailed study	
11	foreign graduate scientist with interest in IK work[1]	J/M	+ BSc	S-M	V – A		FT	a				Culture shock Language Illness
12	foreign postgraduate scientist with interest in IK work	M/S	+ to ++ MSc/PhD	S-D	A – MH	F/P-3/4, >1m	FT	b.c	Some respect for education & technological specialist			
13	foreign postgraduate anthropologist	M/S	+ to ++ MA/PhD	S-D	A – MH	F/P-3/4, >1m	FT	b/c	Anthropology/ social science training Open-minded re IK	No science, irrelevance risk	Will learn language	Steep, long learning curve
14	national postgraduate sociologist	M/S	+ to ++ MA/PhD	S-D	A – MH	F/P-3/4, >1m	FT	a/b		Little influence – respondent probs	Speak language Familiar with culture – rapid entrée	
15	experienced national scientist with interest in IK work	S	+++ PhD	S-C	MH – H	PT-1, <1m	FT	a-c	Technologically knowledgeable Familiar with PRA	Closed-mind re IK, rigid thinker Uneasy with anthro methods		Education may not foster independent research Culture bound
16	experienced foreign scientist with interest in IK work	S	+++ PhD	S-C	MH – H	PT-1, <1m	FT	d/e			Interdisciplinary minded	Language Short visits

No.	Description										
17	experienced foreign anthropologist with science training	S	+++ PhD	S-C	H	PT-1, <1m FT	d/e	Techno-logical & social science experience PRA familiar Senior	Lack of time: multi-task work Limited access to poor	Interdisci-plinary Language poss Know culture – rapid entrée	Prob short visits Rare polymath
18	experienced national social scientist with science training	S	+++ PhD	S-C	H	PT-1, <1m FT	b-d				Culture bound Rare polymath
19	experienced national sociologist	S	+++ PhD	S-C	H	PT-1, <1m FT	b-d	Anthro-pology/ social science Wide experience PRA familiar		Familiar culture – rapid entrée	Culture bound Prob short visits
20	experienced foreign anthropologist	S	+++ PhD	S-C	H	PT-1, <1m FT	d/e			Language (poss. fluent)	Prob short visits, anti-develop-ment

[1] May be VSO-type volunteer on trainee

Table 3.2 Key to Table 3.1

Factor	Code	Explanation
Status (equates with age seniority and respect)	S	Senior
	M	Middle
	J	Junior

Factor	Code	Explanation
Experience (in relation to IK)	+++	10 yrs +
	++	5 years
	+	1 – 2 years
	-	< 1 year

Factor	Code	Explanation
Methods (in relation to ease of use)	S	Simple/easy
	M	Moderately easy
	D	More difficult
	C	Difficult and complex

Output quality (in relation to IK)		Availability (primarily for field work)	FT/PT	Full-time / Part-time	Cost (based on current charge-out rates, not salary)		Day rate
High	H		1/2/3/4 Q/yr	Part-time input is 0–3, 3–6, 6–9, or 9–12 months a year		a	Day rate < £50/day
						b	Day rate £50 to £100/day
Medium-high	MH					c	Day rate £100 to £250/day
			< 1m / > 1m	Time spans of more than or less than 1 month		d	Day rate £250 to £500/day
Average/acceptable	A					e	Day rate > £500/day

Table 3.3   Staff issues pertinent to project design

Staff factor	Issues
IK/anthropology experience	Experience should ensure reliable outputs, but anthropological training may incline towards too much detail or less relevant issues for development project. Able to select most efficient methods. More experience equates with higher costs.
Science/technology experience	IK staff with technological experience are few, but will be able to address more ambitious TORs. More experience equates with higher costs, trade-off increased ability to address objectives.
Local experience	Staff who have previously worked in region will have knowledge of cultural, socio-political etc issues, and more rapidly address project issues. Cost will depend on origin (local, national, foreigner) and qualifications etc.
Seniority	Seniority usually equates with experience and cost. Senior staff may have little available time as busy people.

Consultant	Works only for duration of contract. Focuses on project throughout contract & shares project goals. Institutionally isolated. May tend towards the practical/pragmatic. Consultancy staff usually more expensive than NGO, research institute or university staff but available for the contracted time.
Academic	Long term engagement with IK issues. Many competing demands on time, juggle availability with term-times. (Academic inputs not always short-term, staff may take sabbaticals or charge for replacement teaching staff and focus wholly on project.) Good access to supporting resources (eg research students, library, IT, colleagues, etc) for no extra cost. Academic goals may conflict with project goals. May tend towards the analytical/theoretical.
Gender	Staff may find problems accessing IK held by opposite sex; females may face problems in some societies.
Ethnicity	Staff of different ethnic origin to locals may experience problems. Local staff often cheaper, speak language. But local staff may find it difficult to 'step outside the culture' and make impartial analyses or undertake comparative analysis with other regions.
Language ability	Relates to local experience / ethnicity. Knowledge of language reduces translation costs, speeds up research and may improve reliability of data collected.
The 'I' factor	Ability of staff to interact important for interdisciplinary IK studies. Personal attribute that relates to attitude, dubbed the 'I' factor, significant in selecting staff but difficult to predict.

Table 3.4   Type of indigenous knowledge study appropriate to project duration

3 months Brief	6 months Short	12 months Medium-short	2 years Medium	3 years Extended
Pilot study or research on a specific technology		Increasingly in-depth research ↑ scope and ↑ quality (depth of understanding) with greater duration		

Project cycle management

Time relates to the project cycle, with significant implications for indigenous knowledge research. The cycle that drives projects through defined stages – their completion at a pre-determined rate a measure of management success – may present problems. It is prescriptive. It inclines towards the 'blueprint' approach when such research requires a flexible 'process' one (Nolan 2002: 98–99). Managers should allow for innovations on the project cycle. Participatory research faces the same problem incorporating local people in project design.

The incorporation of indigenous knowledge in project design assumes extensive primary stakeholder involvement. The implication for project cycle management is to allow the time required for in depth indigenous knowledge enquiries; such consultation is ideally a far reaching and slow process. Project managers should lobby for the adoption of longer project cycles, if they think that the trade-offs in the scope/quality of the research are worth it.

Chronology is a key issue. There are three options:

- The preferred approach, when resources and time are sufficient, is to start an indigenous knowledge project cycle before other work, the results, encompassing the local agenda, feeding into the planning phase of a subsequent development project, informing its terms of reference. This arrangement results in a project wave (Figure 3.2), a series of interlinked project cycles that inform and energise one another. The wave flows on and revisits indigenous knowledge as necessary, as new problems emerge; such reiteration recognises that indigenous knowledge is a dynamic resource (this does not lead to endless and unaccountable research; monitoring and evaluation are built into component project cycles).
- An indigenous knowledge component may be introduced elsewhere on the wave. The indigenous knowledge and other research cycles may have to run concurrently rather than in series. This is more likely given current programme management constraints on the wave scenario, which reflect development's demand for quick answers. The indigenous knowledge terms of reference should be loosely defined, allowing for revisions as work proceeds in interaction with other researchers. Ideally some indigenous knowledge work will occur before other research starts, as part of the 'preparation' cycle phase, at the very least as a scoping exercise, as integral to wider project design. Such projects, featuring greater local involvement,

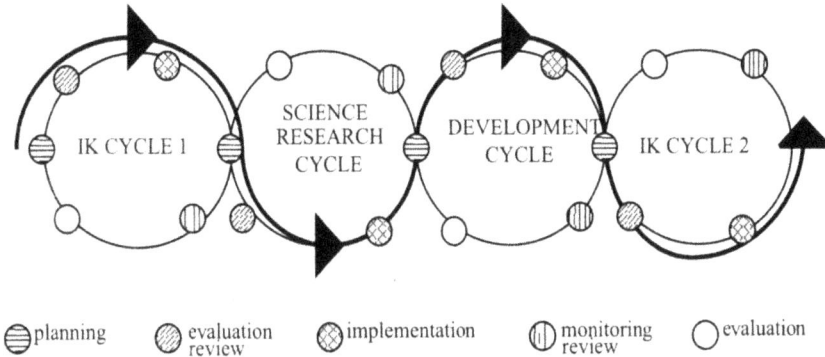

Figure 3.2   **A series of linked project cycles creating a project wave**

will be better focused with interventions more relevant to people. This approach can function, so long as team members exchange information effectively, but synchronisation will demand close attention in such interdisciplinary projects, particularly those attempting to combine the social and natural sciences.

- The performance targets that currently drive research programmes may prevent other researchers waiting for an indigenous knowledge scoping study to report before commencing their work, inhibiting any prior indigenous knowledge investigations. Furthermore scientific research may be needed to focus the indigenous knowledge investigation. If technicians delay their work, the indigenous knowledge research may lack development focus, and proceed as an open-ended ethnographic investigation. Alternatively, where development constraints are thought largely technical, a 'business as usual' project will probably be initiated. Indigenous knowledge may be seen as important only when such a project is underway (when uptake is addressed for example), and an indigenous knowledge study be commissioned as a secondary component. Such 'bolt-on' indigenous knowledge, with locals brought in as collaborators after the start of a project, is not recommended as it is less likely to integrate with other work. It will be brief and limited in scope.

The earlier indigenous knowledge work starts, and the more local input there is, the better the quality of the data. Whichever option you select, you should allow flexibility. The normal review stages of a process project should be part of the terms of reference.

The scope axis

Scope relates to project objectives and links with quality of results. Data reliability is one dimension of research quality, rigour of analysis and depth of understanding are others. In addition to quality, indigenous knowledge work needs to demonstrate 'fitness for purpose' to be incorporated into development. Academic judgements may differ from the demands of development, a practical report being a better fit for some projects' requirements than a scholarly work. Project managers seeking to quantify project scope should ask: "How ambitious is the indigenous knowledge component?"

A project will fall somewhere between three broad levels of scope.

Type (1) projects are of limited scope with simple objectives. They commonly seek basic technical information (e.g. on a single agricultural procedure). You should exercise caution with such narrow studies that threaten to decontextualise knowledge from its socio-cultural context. To treat pieces of indigenous knowledge as transferable and globally relevant information, akin to international scientific knowledge, is to misunderstand and potentially misuse it (Ellen 2000: 169). Such narrowly scoped research often focuses on generating catalogues or databases of local technologies for dissemination. This mining of indigenous knowledge 'nuggets' raises intellectual property concerns. In limited circumstances, Type (1) work can serve a project's ends, but we should recommend employing experienced researchers knowledgeable of the region and culture to reduce decontextualisation dangers.

Type (2) projects are mid-range in scope. They typically feature investigations of defined domains (e.g. crop production, fishing practices or land use). They often adopt an ethno-scientific focus. Several such indigenous knowledge studies have been undertaken in development contexts (e.g. Rhoades 1984; Thorne *et al.* 1997; Lamers & Feil 1995). They have been responsible for major advances in incorporating local knowledge into development to date. They better avoid the decontextualisation pitfalls of Type (1) studies, yet maintain a technical focus with which scientists/development practitioners are able to engage. They are problem-focused studies, and meet the demands of many projects, where Type (3) studies would be overkill, reducing the accessibility of indigenous knowledge and inhibiting its incorporation.

Type (3) projects are ambitious in scope with wide and complex objectives. They are closer to standard anthropological research than the indigenous knowledge work undertaken to-date within the context of development projects. They seek to access understanding of complicated issues set within broad socio-cultural circumstances (e.g. the social, political and ecological dimensions of communal management of a common pool resource). If development continues its move towards tackling more complex problems, such as the Sustainable Livelihoods Approach with its all-encompassing perspective (Carney 1998; Ashley & Carney 1999), Type (3) studies will become more common. Managers need carefully to design projects that incorporate indigenous knowledge work of this complexity, and must resource them adequately and allow them time to undertake the work.

The objective is to make local people's views, ideas and practices accessible to development practitioners/scientists, whatever the scope of a project's indigenous knowledge component. The accessibility demand implies a middleperson 'speaking' on behalf of local people, giving indigenous knowledge data some structure for others (e.g. translating into scientific idioms). In Type (1) studies, the distortion of this translation process is extreme, to the extent that it separates local people and their knowledge. The increasingly participatory, extended ethnographic methods used as we move from Type (2) to Type (3) projects reduces the third party role, empowering people to speak for themselves, lessening translation distortion.

Methods and results

The principal design cube variable regarding methods is the time they demand, which relates to the usefulness and quality of data collected and cost. Managers will trade-off speed against scope, according to the level of indigenous knowledge sought. Methods vary from quick to prolonged. The short-term methods are associated largely with participatory approaches, employed increasingly by those seeking to involve local people in development. The long-term ones come from ethnographic work, employed largely by anthropologists undertaking in-depth field research. Rapid participatory methods are more suited to Type (1) or possibly Type (2) narrower indigenous knowledge studies and ethnographic methods to Type (3) studies of broader scope. Different methods may be used at different times and on a number of occasions throughout the project cycle.

Participatory methods rely heavily on visual techniques, which reduce researcher domination by giving participants more control over information presentation (Craig & Mayo 1995; Chambers 1997). The researcher acts as facilitator. The graphical outputs also foster interdisciplinarity, as readily understandable to those from different disciplinary backgrounds. But the data are shallow. Projects need to balance imprecision against the use they will make of data. These methods suffice to collect technically focused information, but may encourage the seizing of interesting items of indigenous knowledge out of socio-cultural context, leading to possible misinterpretation.

The collection of reliable quality indigenous knowledge generally requires use of longer term, more expensive ethnographic methods, particularly where outputs demand iteration between data collection, analysis and validation. These rely heavily on verbal techniques, the researcher acting as probing investigator extracting and interpreting information (Bernard 1995; LeCompte & Schensul 1999). Ethnographic methods are slow to deliver outputs, mainly due to depth of inquiry. They may deliver results not immediately accessible to others, possibly a detailed and jargon-loaded monograph. While ethnographic context is important, beware of over-collection of data beyond project needs or analysis.

Both participatory and ethnographic methods yield qualitative data largely, not amenable to statistical analysis to assess reliability (except some questionnaire and survey data). They may present problems for scientists used to measuring confidence. The data are difficult for them to assess and use, and they may suspect their validity. The use of rigorous research methods, with frequent triangulation to double check information, should improve faith in such indigenous knowledge data. Together with prolonged engagement with different groups of people, repeated observations, accessing multiple sources of information, investigation of differences, peer and participant checking, parallel investigations and open team discussion.

There is a risk, particularly with short-term research, of dealing with a limited cross-section of a community, often the more wealthy, influential and better educated, who it is easier for outsiders to approach. A balance of informants is necessary, including both genders, all wealth or socio-economic categories, age cohorts, educational levels, occupations, as well as households in different locations and with access to different resources. A socially disaggregated approach, sensitive to economic, environmental, social and institutional issues, is fundamental to understanding the dynamics of knowledge. The longer spent in the field the wider the informant sample and the more representative it will be of the whole community.

Furthermore, the interface between researchers and respondents is not 'neutral' and a project should reflect on its processes of data collection and analysis for biases, making clear the basis on which they collect and interpret data, aware of the 'political' nature of their engagement (Long 2001; Purcell & Onjoro 2002). Researchers should think about the 'interface' between parties to indigenous knowledge research and their 'agendas', much knowledge being negotiated in social encounters. They need to be sensitive to social context, and to the multiple meanings and perspectives that co-exist. Social status and context may influence interactions and data generated, and different groups may vary in their perceptions of issues.

The objective of indigenous knowledge research is an integrated output not a series of parallel final/interim reports on separate disciplinary topics. In order to achieve this it is necessary that an indigenous knowledge informed team works from the start in an 'interdisciplinary' way. The demand for interdisciplinarity should be addressed during the design phase, and related to the project's aims. The indigenous knowledge objectives and outputs should be clear, and how they will interface with technical research results. If any misunderstanding occurs as to what indigenous knowledge research can deliver, this should be addressed early on. If team members continue with wrong ideas, this will result in frustration and disillusionment.

Indigenous knowledge research may produce a range of possible results including: consultancy reports (e.g. Rapid Rural Appraisal); team reports (e.g. Participatory Rural Appraisal); technical reports (e.g. on target technological issue); social situation reports (e.g. Socio-Economic Review); video / film; an interactive ethnographic databank (Barr & Sillitoe 2000); artificial intelligence database; higher degree thesis; and scholarly ethnographic monograph.

A manager needs to match results to a project's aims, and consider how they will integrate with other research and also subsequent development interventions. Key issues are ensuring timeliness of work and synchronisation with other research. There is a danger of indigenous knowledge research falling behind other scheduled work because of the many unforeseen factors that can influence it. A manager needs to be flexible and innovative to keep the indigenous knowledge schedule up with the rest of the project.

The indigenous knowledge design cube: some examples

The budget, time and scope axes need scales to make the cube a useful project design tool. Programme managers will work with values appropriate to their aims.

Time could run from 1 month to 3 years in monthly steps, and resources from £1000 to £500,000 in £1000 increments. We identify only two points on the scale: high and low, for illustrative purposes (more points and the cube combinations become too numerous for presentation). The result is Rubik's cube-like with eight units, each representing a different combination of variables and a distinct indigenous knowledge project. The following sketches outline the management implications for design, staff, methods and results.

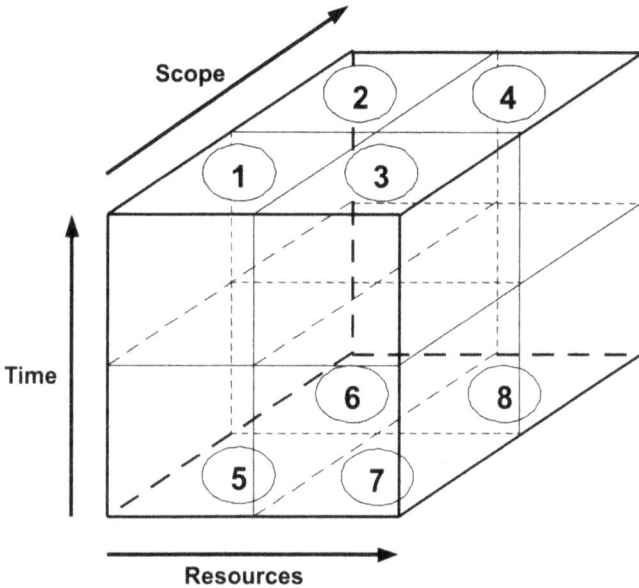

Figure 3.3   Eight indigenous knowledge project examples

Substantial time/limited budget/limited scope

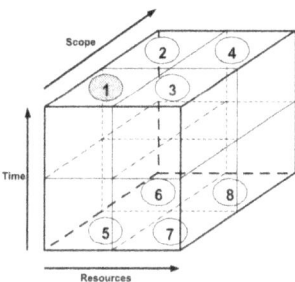

A Type (1) project concerned with a limited technical issue. Although budget is small, <£10,000, there is ample time, possibly two or three years in which to research it. The budget is a key factor, dictating staffing options. Such a project would likely be contracted to a local organisation, e.g. NGO or university department. It may employ a senior person in an advisory role, with a junior team and villagers. There are two risks using entirely local staff, especially a high proportion of village enumerators – ironically, they may be too influenced by the local culture, to which they belong, to act alone as reliable interpreters: They may overlook some knowledge as too obvious to them to be merit recording, whereas such tacit knowledge may be critical to outsiders in deciding project implementation. They may skew the project towards certain local interest groups, notably the wealthier and influential, likely coming under the sway of local power brokers (whose nominees, such as kin, may be proposed for project work), misrepresenting the interests and knowledge of others.

Project managers should use methods to cross-check village enumerator data for local political bias and ensure that they are acting in good faith as gatekeepers between local people and outside agencies. The alternative to contracting a local organisation is to find a senior academic or consultant with previous experience in the locality and knowledge of the technical problem to undertake a rapid appraisal study (although the chances of finding such a person are likely slight) (see cube 5). Methods that yield descriptive data will be appropriate: questionnaires, brief semi-structured interviews, focus group discussions, field observations, participation activities, and photographic documentation. Such projects are unlikely to undertake computer-based data analysis, although expert-system software may be an option if someone locally has the necessary training (such a limited budget project could not afford to train someone). The result is likely to be either a technical report or a catalogue/database of local target technology related to project intervention. There is a danger of such limited indigenous knowledge being taken out of socio-cultural context and inappropriate conclusions being drawn by outsiders.

Substantial time/limited budget/substantial scope

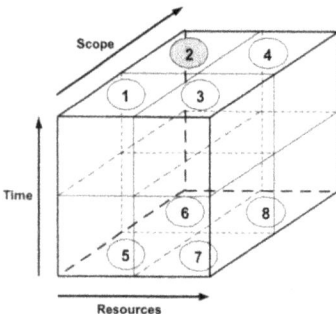

A Type (2) or (3) project of ambitious scope extending over two or three years. The limited budget is again a key factor. This is likely to be a locally staffed project largely, the small budget probably used to best effect employing national NGO staff and village enumerators to collect the large amounts of data sought (note caveats Cube 1). Resources will need to be set aside to train them in varied data collection methods and it is advisable to assign further resources to monitor progress and for external quality assurance. While the project's scope argues for senior staff, it cannot afford such experience, particularly foreign. A national university or NGO sociologist, preferably with some science, could be appropriate. Alternatively a post-graduate researching in the region might give guidance or a more senior academic on a short visit. Methods are an issue given this Cube's scope. Participant observation and open-ended interviews should be used but are beyond such a project. Instead use a range of participatory techniques that junior NGO staff can implement with enumerators, sufficient to tackle the complex indigenous knowledge objectives. Venn diagrams, social mapping, institutional diagrams, wealth ranking, and historical timelines will help understand socio-cultural context. Flow charts, farm walks, mapping, matrices, problem census, participative technology analysis, seasonal calendars, taxonomies, and transects will help access indigenous knowledge domains of interest. Supplement with focus group discussions and interviews, although do not put emphasis on these as analysing interviews on complex subjects is difficult. Computer packages deal with these data well, but are beyond this Cube's resources. It may be possible to experiment with innovative participatory methods, such as community video, given the abundant project time. This will require diversion of limited resources to training and equipment. Participatory video can capture complex local knowledge and empower communities to influence development initiatives. It diminishes the interpreter-gatekeeper role and confronts the issue of researcher subjectivity. It reduces abuse of local peoples' intellectual property rights, unlike more extractive methods that separate owners from representations of their knowledge. Results may be difficult to integrate with other technical research, e.g. PRA reports featuring many techniques and rich with visual/diagrammatic material. Results may include social reviews and community workshop reports, possibly student theses.

Substantial time/substantial budget/limited scope

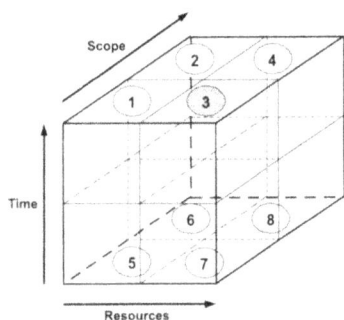

A Type (1) project researching a narrow topic, with ample time (possibly two or three years) and a generous budget. This suggests an imbalanced project, which should be queried, the relevance of the indigenous knowledge to overall aims might be improved by broadening scope. There is a danger of distorted, albeit elaborate outputs, ignoring socio-cultural context with the direction of time and resources towards a narrow technical issue, unless an experienced person, who knows the society intimately, undertakes the work. If the scope remains narrow, the project should employ experienced researchers aware of the wider socio-cultural issues i.e. social scientists with experience of the locality (4, 6, 8, 13, 14, 17–20 Table 3.1). If the indigenous knowledge study and parent project run in parallel for the same period of time, all staff can be involved in the design, implementation and analysis phases, ensuring indigenous knowledge integration. The time available to researchers to interact with other project scientists and return to informants is a bonus. They can present results to the project team, canvas opinion and go back to the field to collect further more targeted data, increasing synergy. Time permits several cycles of data collection, analysis and reflection, to address queries and fill gaps with repeated visits to the field to collect more information. These features of extended duration projects considerably improve the quality and relevance of the data. Such projects will use methods yielding largely descriptive information on local technology, how employed, people's ideas about it etc.; and may include questionnaires, short semi-structured interviews, focus group discussions, field observations, farm walks, and participant observation. The resulting catalogue documenting the local technology may be ambitious, even publishable, given available resources, and include illustrations and photographs, even documentary video of things in use. Resources allow for use of computer databases, although the narrow range of data may not repay the time invested coding it. The descriptive data are amenable to expert system unitary statement documentation, although the knowledge base will be limited and yield potentially distorted outputs.

Substantial time/substantial budget/substantial scope

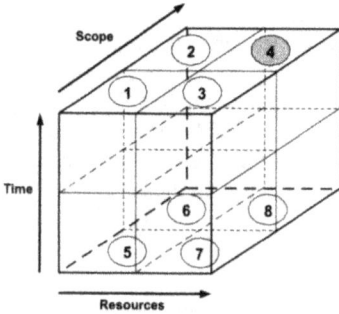

A Type (3) project of broad scope, well funded (£100,000+) and of long duration, the ideal for high quality indigenous knowledge work. The resources indicate esteem for indigenous knowledge. Such a project may be implemented as the first cycle in a 'project wave'. It may build a team of senior and junior staff from national and foreign institutions. It can afford to employ experienced staff for extended periods, who can supervise a range of methods to collect data of high quality and reliability. The project might employ an experienced sociologist or anthropologist as co-ordinator and an experienced scientist with indigenous knowledge interests as a senior partner, supervising a team of field-level staff comprising NGO staff with social science experience, local social science graduates or post-graduates, and village enumerators trained in indigenous knowledge work, and possibly Masters/PhD students. The building and management of such a complex team will be an important task, overseeing information flows both within it and beyond to the wider project. There is the time to promote interaction between team members and develop an effective communication strategy between staff to ensure information exchanges. Such a mix of staff can use a wide range of methods, requiring both short and extended local contact to engage with the full complexity of knowledge in socio-cultural context. All methods are potentially available including extensive use of interviews and focus group discussions on complex subjects, having the resources to analyse such data. The prolonged involvement of senior staff makes ethnographic methods such as open-ended interviewing and participant observation also available. Such a project should yield highly reliable data and in depth understanding of issues with prolonged interaction with local people. It might also reap the benefits of computer-based data analysis and expert systems software being able to employ trained staff for the necessary time. The project may produce a range of results. It may help draft the terms of reference for a subsequent partner technical project. Both projects should aim to contribute to an integrated final report on the issue jointly researched, incorporating relevant local knowledge and socio-cultural context. The project may supply social situation reports (e.g. a socio-economic review of the target community). It may produce detailed reports on particular topics that yield notable indigenous knowledge (e.g. on soil management or pest control). There may be the resources to pioneer some of the new people-first data collection methods, such as participatory video. It may establish an indigenous knowledge database, integrated into wider project databases, perhaps depositing in country for on-going data entry by local researchers. In addition to project specific outputs, there may be more academically oriented ones, such as higher degree theses (although a drawback is that these may not be available until after the project when candidates complete), and, with extended involvement of senior university-based staff, scholarly ethnographic monographs and papers in peer-reviewed journals, some of which may appear during the project, others later when the researchers have had time to reflect on their findings.

Limited time/limited budget/limited scope

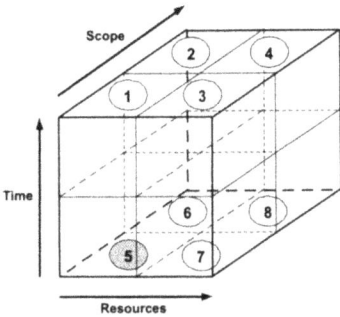

A Type (1) project working on a limited technical topic, with little time (3 to 6 months), and a small budget (<£10,000). Such a mini indigenous knowledge project suggests either a limited study commissioned to complement a project from the beginning, possibly in a 'tick a box' way; or an emergency study commissioned during a project when it is realised that some understanding of indigenous knowledge is necessary, maybe a 'fire-fighting' exercise; or that the person who has conceived the project does not understand what indigenous knowledge research entails. It is probable that scientists are driving such a limited scope project unaware of the weaknesses of socio-cultural decontextualisation, and oblivious to the shallow understanding and dubious data with the time and resources only to skim the subject. It is necessary to involve indigenous knowledge staff at the project design phase to prevent the occurrence of such potentially flawed projects. Indigenous knowledge considerations should feature at the project identification and preparation stages to synchronise with the wider project. Such projects with limited budget and time have to make constrained 'either-or' choices about staff and methods. They may be contracted to a local NGO or a university department, with the caveats that apply to local/village staffed only projects. Alternatively, the budget may cover an experienced academic/consultant for about one month. The limited time suggests someone familiar with the region and subject matter, if such a person can be found. Methods will be those that supply descriptive data largely, as for Cube 1. The result may be either a technical report on the target issue (if a consultant employed) or a catalogue/database of relevant technologies (where a local organisation used). The limitations of these outputs need to be borne in mind, as for Cubes 1 and 3.

Limited time/limited budget/substantial scope

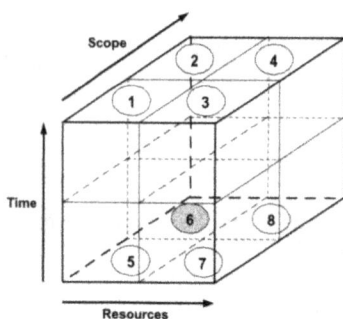

An ambitious Type (2/3) project of short duration and small budget. An unrealistic combination, unless able to draw on substantial research previously undertaken in region on problem. Such a project will be seeking to recruit a senior researcher (17–20, Table 3.1) with experience of locality and issues. If you come up with such a project using the decision cube, you may be well advised to think again. The chances of finding suitably experienced researchers are slight and without such persons the work envisioned is impracticable. Review project requirements and either narrow the scope or increase available resources for the indigenous knowledge work. If it goes ahead, such a project can only afford a few months of staff time. It will rely heavily on secondary sources, possibly with some limited field work to fill in gaps. Methods may include some selective interviewing and use of PRA tools to supplement previously collected data. The result will be an indigenous knowledge consultancy report featuring some ethnographic analysis of issues. There will be little time for interaction with other project researchers, and the indigenous knowledge report will likely be independent of them.

Limited time/substantial budget/limited scope

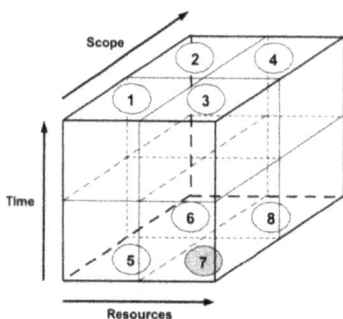

A Type (1) project of narrow scope, to be completed in brief time frame (3 to 6 months), with a generous budget (>£100,000). Such a project is possible where it is realised that some knowledge of local issues needed in a large on-going project and circumstances demand it be short-term and targeted. The project manager appreciates that indigenous knowledge studies are not necessarily cheap exercises, and has the resources available to fund the work properly, although may not fully realise what it entails regarding socio-cultural context. The budget could cover considerable input from senior/consultant staff, but a mixed team of senior supervisory staff and junior field staff might prove appropriate. Opportunities for post-graduate researchers are limited given the short timetable. While staff need not have local experience due to the narrow project focus, it would be advantageous if they have some, for they will be better placed to put the results in socio-cultural context. Methods will focus on collecting descriptive data, as in Cubes 1, 3 and 5. Probably yielding a technically oriented, stand-alone indigenous knowledge report. This may be a technical account or a catalogue/ database of local technologies. There are resources to commission more expensive outputs such as video documentation. The indigenous knowledge work is unlikely to integrate significantly with other research, probably being out of synchronisation with it, the short time frame also reducing opportunities for interaction with others.

Limited time/substantial budget/substantial scope

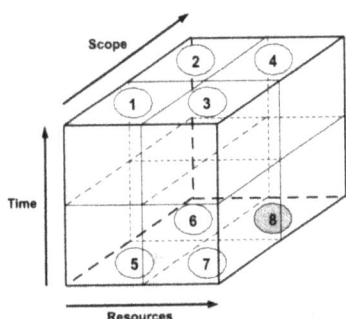

A Type (2/3) project of ambitious scope, undertaken in a short period (3–12 months) with a generous budget (>£100,000). The key design issue is the imbalance between broad scope and short time scale. The project requires an in-depth engagement with indigenous knowledge in socio-cultural context, which demands a extensive interaction with local people. The brief time available is inadequate. The mismatch between scope and project duration can only be solved if staff already have an in-depth appreciation of the local culture, having worked in the region previously. The solution is to employ an expert who can meet the high expectations in a short period. Hire the best person in the field that you can afford. The budget could employ a post-doctoral researcher for 3–4 years, or professor/ consultant for $1-1\frac{1}{2}$ years. A team is unlikely to gel in the short time available (unless recruited as an established institutional/ consultancy group, which is rare), although a small local support team might increase efficiency. The field of qualified persons for such a project is limited. They should not only be field-experienced anthropologists or rural sociologists, but also persons sympathetic to the demands of development, as the indigenous knowledge study, working to such a tight deadline, will probably be running in parallel to an on-going parent project that has an urgent need of its information. Time will constrain the methods used. Careful selection of key informants by experienced staff and skilful interviewing will probably be central. Researchers may supplement this with other ethnographic techniques and judicious use of PRA methods, possibly including social mapping, institutional diagrams, seasonal calendars etc. Results are likely to be ethnographically detailed consultancy-type reports; also possibly social situation reports on target communities. They will probably be independent since the project will be too short and demanding to build interdisciplinary team linkages. There will be insufficient local participation to produce a participatory output. And insufficient time to code and analyse material using computer databases.

Managing projects with indigenous knowledge component

It is routine for teams to conduct development research. Interaction between natural and social sciences is fundamental. Success depends heavily on how well the team functions, members needing to work effectively together to realise the potential insights and synergies of indigenous knowledge research. There are basically two approaches. Multidisciplinary, where researchers work in parallel and largely independently, finally contributing sections to a report. And interdisciplinary, where researchers work together from the start and try to integrate their research throughout the project (Cochrane 1976; Chubin, Porter & Rossinin 1986). The terms of reference will need to specify which, as it impacts significantly on the design and implementation of indigenous knowledge research. A multidisciplinary project (cubes 5, 7, 8) will commission a separate indigenous knowledge study to link with other work at an advanced stage, resulting in an unintegrated final report. Indigenous knowledge will have little impact on other research. An interdisciplinary project (cubes 2, 3, 4) will commission indigenous knowledge as an integrated part of other research, continually iterating knowledge with other team members. It is potentially more insightful, and certainly more ambitious in management terms, as the project must be established and managed to facilitate such interaction.

The management of a team, particularly an interdisciplinary indigenous knowledge one, presents many challenges in both design and implementation phases (Sutherland & Martin 1999). Multi-interdisciplinary indigenous knowledge team research featuring diverse subjects in foreign cultural contexts demands a wide range of intellectual capabilities. It is not 'textbook' routine, and frequently requires interpretation of new knowledge. It demands flexible teams capable of innovation and rapid re-evaluation of goals, which presupposes necessary intellectual calibre, experience and skill. It requires expert management to integrate natural and social science investigations – broadly quantitative and qualitative respectively. The administrative costs are high ensuring information passes between different disciplines, exacerbated if researchers are dispersed in different institutions – such teams are usually ad hoc, and demand bonding before effective cross-disciplinary research can start (Sillitoe 2004).

Beware of a team comprising several specialists only, which may be unable to integrate and work in an interdisciplinary mode. Smaller teams with a stable membership integrate best, assuming members are compatible. A successful team will comprise a mix of personalities that complement one another in promoting innovative work (Belbin 1981). Comradeship, interpersonal relations and interdependence are ingredients necessary to success. We often overlook these in recruiting research teams, evaluating individuals' qualifications and disciplinary skills, but not necessarily their potential contribution to team social chemistry. Allow realistic time for team recruitment and bonding– which current development funding arrangements discourage with their politically driven demands for immediate action. Ability to bond is largely a personal attribute (the 'I' factor, Table 3.3) that is difficult to predict. It is not something assessable from CVs; it requires a trial period in a team. Someone may fit into one team well but be a disaster in another.

A team is likely to experience some conflict initially before 'normalising', at which point it can start to realise the benefits of team working (Handy 1985). Some tension between disciplines will be the rule rather than the exception, indeed some intellectual tension is necessary for creative research. In indigenous knowledge projects this 'forming' and 'storming' stage can be crucial to sort out professional differences between team members. The absence of shared disciplinary perspectives is acute in teams that bridge the social-natural science boundary. Part of the 'norming' stage has to be establishing clear lines of communication between team members across this boundary. These will comprise the bridge that links the team together, necessary if 'performing' is not to disappear down the disciplinary gulf.

A collegiate approach to indigenous knowledge research is recommended, which promotes equal interaction between researchers. This peer-based approach may feature a co-ordinator to facilitate research interaction. This role involves less direction than encouragement of creative co-operation, and an ability to communicate, criticise constructively and resolve any conflict. Share responsibilities to foster joint ownership of project and avoid co-ordinator overload. Plan and agree goals as a team activity. The 'integration by leader' approach is not recommended. It is likely where staff/consultants are on short-term contracts. It places heavy pressures on the leader. It is intellectually difficult to achieve, demanding a polymath – a rare person in today's highly specialised research world. Establish procedures to ensure good communication, drawing the team together through constant interaction and exchange of ideas to promote integrated analysis.

Note

1   We developed this model in projects generously funded by the Natural Resources Systems Programme of the DFID, under contracts R6744 and R6756. The views expressed are those of the authors and not necessarily DFID.

References

Ahmed, Z.U. 2000. When a Bangladeshi 'native' is not a Bangladeshi 'native'. In *Indigenous Knowledge Development in Bangladesh. Present and Future.* (ed.) P. Sillitoe. London:Intermediate Technology Publications & Dhaka: University Press. 203–209.

Alam, M. 2000. Indigenous knowledge fieldwork: Interaction with natural resource scientists. In: Sillitoe, P. (ed.) *Indigenous Knowledge Development in Bangladesh. Present and Future* London: Intermediate Technology Publications & Dhaka: University Press. 197–202.

Ashley, C. & Carney, D. 1999. *Sustainable livelihoods: Lessons from early experience.* London: Department for International Development.

Barr, J.J.F. & P. Sillitoe. 2000. Databases, indigenous knowledge and interdisciplinary research. In *Indigenous Knowledge Development in Bangladesh. Present and Future.* (ed.) P. Sillitoe. London: Intermediate Technology Publications & Dhaka: University Press. 179–195.

Belbin, M. 1981. *Management Teams – why they succeed or fail.* London: Butterworth Heinemann.

Bernard, H.R., 1995. *Research methods in anthropology.* London: Sage.

Brush, S. & Stabinsky, D. 1996. *Valuing local knowledge: Indigenous people and intellectual property rights.* Washington DC.: Island Press.

Carney, D. (ed.) 1998. *Sustainable rural livelihoods: What contribution can we make?* London: Department for International Development (papers presented at International Development Advisers' Conference July 1998).

Chambers, R. 1997. *Whose reality counts? Putting the first last.* London: Intermediate Technology Publications.

Chubin, D.E., Porter, A.L. & Rossinin, F.A. 1986. *Interdisciplinary analysis and research. theory and practice of problem focused research and development.* Mt. Airy: Lomond.

Cochrane, G. (ed.) 1976. *What we can do for each other. An interdisciplinary approach to development anthropology.* Amsterdam: B.R. Grüner Publishing Co.

Craig, G. & Mayo, M. (eds.) 1995. *Community empowerment: A reader in participation and development.* London: Zed Books.

den Biggelaar, C. 1991. Farming systems development: Synthesizing indigenous and scientific knowledge systems. *Agriculture and Human Values* 8 (1&2): 25–36.

Ellen, R.F. 2000. Local knowledge and sustainable development in developing countries. In *Global sustainable development in the 21st century* (eds.) K. Lee, A. Holland & D. McNeill. Edinburgh: Edinburgh Press. 163–186.

Emery, A.R. 2000. *Integrating indigenous knowledge in project planning and implementation.* Nepean (Ontario): Partnership Publication with Kivu Nature Inc., The International Labour Organization, The World Bank and Canadian International Development Agency.

Grenier, L. 1998. *Working with indigenous knowledge. A guide for researchers.* Ottawa:IDRC

Gustafsson, R. 1995 *The way we work with indigenous and tribal peoples.* INDISCO Guidelines for Extension Workers, No. 1. Geneva: Cooperative Branch, International Labour Office.

Handy, C. 1985. *Understanding organisations.* Harmondsworth: Penguin Books Ltd.

IIRR 1996. *Recording and using indigenous knowledge: A manual.* Silang, Cavite, Philippines: REPPIKA, International Institute of Rural Reconstruction.

Laird, S. (ed.) 2002. *Biodiversity and traditional knowledge.* London: Earthscan.

Lamers, J.P.A. & Feil, P.R. 1995. Farmers' knowledge and management of spatial soil and crop growth variability in Niger, West Africa. *Netherlands Journal of Agricultural Science* 43: 375–389.

LeCompte, M.D. & Schensul, J.J. 1999. *Designing and conducting ethnographic research.* Volume 1 Ethnographer's Toolkit Series. Walnut Creek, CA: Alta Mira Press.

Long, N. 2001. *Development sociology: Actor perspectives.* London: Routledge.

Nolan, R. 2002. *Development anthropology: encounters in the real world.* Boulder: Westview.

O.D.A. 1988. *Appraisal of projects in developing countries.* London: HMSO Books.

Purcell, T. & Onjoro, E.A. 2002. Indigenous knowledge, power and parity: models of knowledge integration. In *'Participating In Development': Approaches To Indigenous Knowledge* P. Sillitoe, A. Bicker, and J. Pottier (eds.) London: Routledge (ASA Monograph No. 39). 162–188.

Rajasekaran, Bhakthavatsalam. 1994. *A Framework for incorporating indigenous knowledge systems into agricultural research, extension and NGOs for sustainable agricultural development.* Studies in Technology and Social Change, No. 22. Ames: CIKARD, Iowa State University.

Rhoades, R.E. 1984. *Breaking new ground: Agricultural anthropology.* Lima International Potato Centre, Lima, Peru.

Sillitoe, P. 1998. The development of indigenous knowledge: a new applied anthropology *Current Anthropology* 39 (2): 223–252.

Sillitoe, P. 2000. Cultivating indigenous knowledge on Bangladeshi soil: an essay in definition. In *Indigenous knowledge development in Bangladesh: Present and future.* P. Sillitoe (ed.) London: Intermediate Technology Publications & Dhaka: University Press. 145–160.

Sillitoe, P. 2002a. Participant observation to participatory development: Making anthropology work. In *'Participating In Development': Approaches To Indigenous Knowledge.* P. Sillitoe, A. Bicker & J. Pottier (eds.) London: Routledge (ASA Monograph No. 39). 1–23.

Sillitoe, P. 2002b. Globalizing indigenous knowledge. In *'Participating In Development': Approaches To Indigenous Knowledge.* P. Sillitoe, A. Bicker & J. Pottier (eds.) London: Routledge (ASA Monograph Series No. 39). 108–138.

Sillitoe, P. 2004. Interdisciplinary experiences: Working with indigenous knowledge in development. *Interdisciplinary Science Reviews* 29 (1): 1–18.

Sillitoe, P. Bicker, A. & Pottier, J. (eds.) 2002. *'Participating In Development': Approaches To Indigenous Knowledge.* London: Routledge (ASA Monograph Series No. 39).

Sutherland, A. & Martin, A.M. 1999. Institutionalising farmer participatory research – key decisions based on lessons from Africa. In *Decision Tools for Sustainable Development,* I.F. Grant and C. Sear (eds.) Natural Resources Institute, University of Greenwich, Chatham Maritime. 46–65.

Thorne, P.J., Sinclair, F.L., & Walker, D.H. 1997. Using local knowledge of the feeding value of tree fodder to predict the outcomes of different supplementation strategies. *Agroforestry Forum* 8 (2): 45–49.

Warren, D.M. 1991. *Using indigenous knowledge in agricultural development.* World Bank Discussion Papers, No. 127. Washington, D.C.: The World Bank.

Chapter 4

Triangulation with Técnicos: A Method for Rapid Assessment of Local Knowledge

Jeffery W. Bentley, Eric Boa, Percy Vilca and John Stonehouse

Framing the scene

Putting a camera on a tripod can help a photographer get a clear picture. In the same way, seeing a set of knowledge through the eyes of farmers, scientists *and* a third group – *técnicos* – can help get a clearer, quicker picture of local agriculture. The knowledge of técnicos provides a perspective, which can be compared with farmer and scientific knowledge to offer new methods for rapid inventories of local knowledge.

By scientists, we mean people with advanced degrees, who conduct research, especially in a laboratory or on a research station. They tend to be senior people, based in a city. It is important to distinguish scientists from another group, known in Spanish-speaking America as 'técnicos' (technical people). Técnicos are an occupational group in their own right.

In the past 20 or 40 years, técnicos have emerged in many countries as a grassroots, occupational group involved in all of agriculture, especially in the formal sector. They may be extensionists, or agronomists on an isolated station, or even managers of credit programmes. Técnicos are often posted to small towns. Some técnicos write well and publish (e.g. see Padilla 1999 for an exemplary booklet on peach trees). But many técnicos are not rewarded for publishing, and writing is not always their favourite task. So técnicos tend to publish less than scientists and perhaps for this reason have been somewhat overlooked in the formal development literature.

Characteristics

In this chapter, we define técnicos as people with formal education in agriculture or a technical field, but not with master's or doctorate degrees. Técnicos almost always work for an institution. Técnicos frequently come from a rural background. Many are the sons or daughters of farmers. They often live in isolated communities, while their children attend schools in larger towns, so the técnico is away from home a lot. Técnicos are often bi- or multi- lingual. They talk to farmers in a local language, such as Quechua in the Andes, or Malayalam and Kannada in Southern India, and speak with scientists in an international language such as Spanish or English. Some técnicos are also farmers, working a bit of land as an income supplement, to rear improved seed, for the love of farming or for all of the above.

Técnicos may have less international experience than the scientists, but often have much more local field experience than the scientists. There are técnicos who indulge in stereotypical behaviour: sitting in their jeep and honking the horn for farmers to come to the car. But many técnicos are more sympathetic: driving and walking into remote villages, eating with farmers, and playing cards or football with them, besides conducting technical work. In other words, técnicos are usually socially skilled and comfortable with rural people. But técnicos are not altruists. They are doing a job, not applying for sainthood.[1]

Agricultural and social scientists visiting rural communities frequently work through the good graces of técnicos, who help us meet and interview farmers. Visiting scientists on farmer surveys sometimes see técnicos as a necessary evil, a kind of watered-down researcher. The agricultural or social scientists may become frustrated when técnicos interrupt and contradict the farmers during interviews.

We believe that técnicos have a more constructive, proactive role to play in farm surveys, and that técnicos interrupt farmers because they feel frustrated at not being listened to. At the end of this paper we suggest a method that helps involve the técnico in research of farmer knowledge.

Information brokers

Técnicos are often employed as extensionists, teaching science to farmers. But técnicos can be just as good explaining farmers' ideas to scientists. For example, one of us (JB) recently heard a técnico in India explain why farmers could not adopt a researcher's recommendation of removing early coffee blossoms to control the coffee berry borer. The técnico demonstrated by picking off a few early flowers, explaining that as he did so, he also damaged the buds that would grow into the berries of the main harvest.

Many scientists do not appreciate the potential of técnicos to convey farmer information to scientists. Government officials in Paraguay once asked a visiting anthropologist (JB) why Paraguayan farmers planted cotton at the traditional spacing, instead of adopting denser spacing. The anthropologist, who had just arrived in Paraguay, did not know. The room was full of Paraguayan extension agents who had worked for years with cotton farmers, but the scientists did not ask their opinion. Later, these técnicos explained to the anthropologist that the farmers planted cotton at wide distances so they could weed with horse-drawn tools. The farmers were constrained by labour, not by land, and so were not motivated to plant densely, but did need to weed on time.

Language

Técnicos have a vocabulary of terms that we call common-technical. These names can bridge scientific and folk names. Definitions are as follows:

Scientific names: Of neo-Latin[2] etymology. Formal and decided by internationally agreed codes, which are periodically reviewed by taxonomists. Scientific names are binomials that label organisms to the species level. Scientific names are organised in nested hierarchies (species, genera, families, orders...). The genus name is sometimes used to describe a disease, as in phytophthora root rot of pepper.

Folk names: Used by farmers and other rural people to describe things in the real world. (Folk names are also called common, local and indigenous names. In Spanish they are often called 'popular'.) Etymologically, folk names are from local languages and the categories are organised in hierarchical, folk taxonomies (Berlin 1992, Ellen 1993, Bentley & Rodríguez 2001). In the case of diseases, they usually label symptoms or the disease condition, more often than the causal agent, hence *cacar daun* in Indonesia meaning 'leaf pox'. And *k'aspara* in Quechua (from the verb k'aspay, 'to singe') refers to discoloured leaves, regardless of the cause).

Common-technical names: Used by técnicos. Some are loaned from the Linnaean lexicon and may include older scientific names that taxonomists have replaced. The scientific name for the fungus of frosty pod disease of cacao is *Moniliophthora*, yet técnicos call it by an earlier scientific name, Monilia. Other common-technical names come from local languages. For example, Latin American técnicos have disease words like *torque, tizón, tristeza*, which are from standard Spanish. Others are from Native American language, for example coñera[3] for certain animal diseases.

Some popular names are loaned from agronomy literature in English (e.g. 'damping-off'). Others are loan translations (calques) from English: palomilla dorso de diamante is a Spanish translation of 'diamondback moth'. Like folk names, the common-technical names used by técnicos are problem-oriented: they group pests by the damage they cause (e.g. the Andean potato weevil is at least 14 insect species in 3 genera of Curculionidae). Like scientific names, common-technical names pay attention to biological integrity, e.g. calling larvae and adult insects by the same label, while folk taxonomies often have separate names for the caterpillars and moths of the same species (Hunn 1982).

Bridging the gap

The common-technical vocabulary of técnicos is relatively stable through time and over large geographical areas.[4] Técnicos who took part in a project on tree health in Bolivia were from different areas in the department of Santa Cruz, yet used the same common names (Table 4.1). These técnicos knew the folk names that smallholder farmers used. The técnicos tended to have a single, unambiguous common name for each phenomenon, which helped us to sort out the various regional synonyms of the folk names. The técnicos also helped us understand the multiple meanings of some folk names. For example, farmers use the term musuru to describe not only peach leaf curl, a disease caused by the fungus Taphrina deformans, but also aphid damage. Both organisms curl leaves, but in a different way.

Table 4.1 shows the correspondence between one set of folk names and common-technical names for diseases in Bolivia.

Table 4.1   Names of peach diseases in Bolivia, as reported by técnicos

Common-technical name		Bolivian folk name [Spanish names are in italics, Quechua names in bold]		Scientific name
Spanish	**English**	**Spanish and Quechua**	**English translation**	
Mosca de la fruta	Fruit fly Mediterranean fruit fly	*gusano, tábano,mosc a de la fruta,* **pilpintu**[5]	worm (caterpillar) 'tabanid' fruit fly moth/ butterfly	Insect: possibly *Anastrepha* sp. (Diptera: Tephritidae) or *Ceratitis capitata* (Diptera: Tephritidae)
Monilia	Monilia	*pasmo pudrición monilia sarna corcho*	cold rot monilia mange cork	Fungus: *Monilinia* spp.
Torque	Peach leaf curl	**musuru** *torque* *somorotota*	Original meaning of musuru was maize smut (*Ustilago maydis*) torque is from standard Spanish	Fungus: *Taphrina deformans*
Cochinilla	Scale insect	*cochinilla ceniza escama piojo*	cochineal ash scale louse	Scale Insect: not identified
Pulgón	Peach aphid	**musuru** *pulgón pulguilla piojo*[6]	See above big flea (i.e. aphid) little flea louse	Aphid: *Myzus persicae?*
Barrenador	Borer	*taladrillo*	little drill	Insect: not identified
Oidium	Mildew	*cenicilla polvillo*	little ash powder, dust	Fungus: not identified

A methodological suggestion

Twenty years after Robert Chambers (1980) chided development professionals for working like 'tourists', there is still a need for reconnaissance methods that are fast enough to give timely feedback, while providing more than a tourist's depth of understanding. We suggest the following method for starting to elicit local knowledge. It builds on the strength of local técnicos, who are definitely not development tourists; they have years of experience in a region, and can share that knowledge in a way that scientists understand. We have found that it works well for a quick inventory of farmer categories of pests and diseases, and it may have other applications.

Pests and diseases are difficult to study in the field; many are only present in the rainy season, when backcountry travel is the most difficult (Van Mele 2000). The names farmers use for these organisms may be quite local, and a good definition of them must take into account a scientific identification of the organism(s), its stage of life when it causes its typical damage (e.g. larva), the part of the plant infested etc. Técnicos can articulate these complex definitions, even when the pests are temporarily absent. Técnicos can explain local knowledge verbally, or in writing, or in group exercises with markers and butcher paper, using the following method.

Method

1. Meet the técnicos and discuss their work with them. Use semi-structured interviews, techniques from PRA, or just talk.
2. Find out what the técnicos think about the problem. For example, the topic can be 'What are the main pests of fruit trees?' The problem might concern human health or livestock.
3. Ask the técnico what the local people know about these topics (e.g. local, folk definitions of pests and diseases).
4. Organise this material as hypotheses, or things to confirm in the field. The hypotheses can be written, or a mental checklist.
5. Select topics to discuss with rural people, based on these hypotheses. Examples: pests and diseases of fruit trees and their importance; or pests and diseases of potatoes, their aetiology and control.
6. Hold semi-structured interviews with farmers, starting with these topics.
7. Encourage the técnico to ask the questions. Most técnicos are comfortable asking farmers questions, and the farmers will understand the técnico better than the visiting scientist. Giving the técnico an active role in the interview will keep him or her from interrupting and contradicting farmers, and will give the visiting scientist more time to listen and take notes.

Case study

In August, 2001, Bentley and Vilca conducted a study of local knowledge of potato pests and diseases in Peru, for the International Potato Centre (CIP). The study was part of a DFID-funded project on bacterial wilt and other diseases. We needed to understand local categories of potato problems in Huánuco (Central Andes) and Cajamarca (Northern Andes). We began each study by meeting with local agricultural scientists, extensionists and other technical people. We asked them to list the local folk names for pests and diseases of potato. The técnicos in Cajamarca made the following list in about 10 minutes (see Table 4.2).

Table 4.2 Local names of potato pests, listed by técnicos, Cajamarca, Peru

Rancha	Gusanera
Bacteria	La pus (la lechera)
Cáncer	Liendrecilla
Marchitez	Zorrillo
Moco	Seca seca
Pudrición	Chamso
Rancha amarilla (polilla)	Huyo
Candelilla	Gusano blanco
Mosquilla	Gusano arroz
Vaquita	Ryzoctonia
Shipe	Tictes
Gusano de tierra	–
(Técnicos in Huánuco made a longer list, not shown here).	

We then asked the técnicos to write short definitions for the terms. The people responded with good, short definitions, that included criteria such as scientific names, life stage of the pest, symptoms, and part of the plant affected. The definitions were sensitive reflections of folk categories, e.g. occasionally lumping together fairly different organisms. For example, the definition of '*gusano de tierra*' included larvae of 2 genera of Lepidoptera and larvae of one family of Coleoptera (see Tables 4.3 and 4.4), which reflects the way local farmers use the term gusano de tierra.

Table 4.3 Local folk categories of disease, described by técnicos

Folk Name	Etymology[7]	Meaning
Rancha	?	Disease of the stalk, leaves and tubers, caused by the fungus *Phytophthora infestans*
Bacteria	Borrowed from technical vocabulary	Wilt and black leg (*pierna negra*), caused by bacteria *Ralstonia solanacearum, Erwinia* sp.
Cáncer	Borrowed from technical vocabulary	Wilting and death of the plant, caused by *Ralstonia solanacearum*

Marchitez	Standard Spanish for 'wilt'	Dryness and falling by the plant, with secretions from the tuber, caused by *Ralstonia solanacearum*
Liendrecillo	From standard Spanish, 'liendre,' meaning nits (lice eggs) for the shape of the nematode cysts on potato roots	Small white balls on the roots, caused by the nematode *Globodera* sp.
Pudrición	Standard Spanish 'rot'	Spoiled, rotting tuber, caused by bacteria: *Erwinia* sp., or fungi: *Pythium* spp., *Fusarium* spp., etc.
Seca seca	From standard Spanish 'seca' meaning 'dry'	Drying of leaves and stalks, caused by *Phytophthora infestans*
La pus	From standard Spanish, meaning 'pus'	Secretion caused by the tuber infested with bacterial wilt
Ticte	?	Black crust on the tubers, caused by *Ralstonia solanacearum*)
1. Virosis 2. Plantas cansadas	1. From technical vocabulary. 2. 'Tired plants'	Mosaics, leaf rolling, APMV
1. Plantas amarillas 2. Virosis 3. Hoja de lima	1. 'Yellow plants' 2. From technical vocabulary 3. 'Lime leaf?'	Potato yellow vein virus (PYVV)

Table 4.4 Folk categories of arthropods (and a vertebrate) as described by técnicos

Folk Name	Etymology	Meaning
Rancha amarilla	'Yellow rancha'	Leaf yellowing caused by the larvae of *Phthorimaea operculella* (Lep: Gelechiidae) boring into the plant stalk
Candelilla	'Little candle'	Adult of the moth *Symmetrischema tangolis* (Lep: Gelechiidae) in potato stores
Polilla	Standard Spanish for 'moth'	Damage by the larva of the moth *Phthorimaea operculella* in potato stores

Gusano de tierra	Worm of the earth	Caterpillar in the soil, which eats the tubers, *Spodoptera* spp., *Agrotis* spp. (Lep: Noctuidae) & Elateridae (Coleoptera)
Huyo	?	Cutworm of small plants: *Spodoptera* spp. & *Copitarsia* spp. (Lep: Noctuidae)
Vaquita	Little cow	Green, leaf-eating insect, *Diabrotica* spp. (Coleoptera: Chyrsomelidae)
Mosquilla	'Little fly'	Adults of *Epitrix* spp. (Coleoptera: Chrysomelidae), which make holes in leaves
Zorrillo	Standard Spanish for 'skunk'	Skunks dig up the tubers, which turn green or are damaged by birds
Shipe	From Quechua?	Adults of *Epitrix* spp. & *Diabrotica* spp. (Coleoptera: Chrysomelidae) which eat stems and leaves
Chamson	?	White grub *Buthinus marmon* (Coleoptera: Scarabeidae)
Gusanera	'Mass of worms'	Tuber damage by the Andean potato weevil
Gusano blanco Gusano arroz	'White worm' 'Rice worm' (referring to the colour and shape of the worm)	Larvae of the Andean potato weevil (Coleoptera: Curculionidae)

During the next 4 days after each meeting, Bentley, Vilca and some técnicos visited farmers in their fields. In and around Cajamarca we went to 17 farms in 3 provinces (San Marcos and Cajabamba in Cajamarca, and Huamachuco in the neighbouring department of La Libertad). In structured interviews, farmers told us of the following categories of pests and diseases:

Table 4.5 Diseases, according to farmers

Folk name	Meaning	Identification[8]	No. of times mentioned	Observations by farmers
Rancha, Rancha negra, Hielo fungoso, Quemazón La plaga	Rancha, Black rancha, Fungal ice, Burning, The pest	*Phytophthora infestans*	8	'Sometimes rancha takes the whole crop.' 'The plant turns yellow. It looks like it is ripening and then it is ruined. (The improved potato variety) Canchán has a lot of it.'
Marchitez	Wilt	Bacterial wilt	7	'Wilt is a disease which totally destroys and we do not know how to control it.' 'The potato rots and pus comes from its eyes.'
Amarillas Virosis	Yellow ones Virus disease	PYVV virus	6	'The plant turns yellow. The whole leaves turns yellow. When it has fruit (tubers) it turns green again and grows a little fruit.'
Hielo, Granizada	Ice, Hail	Cold, frost, hail	4	'Frost does the most damage. In mid summer a frost falls, and everything turns white and frozen.'
Mancha negra	Black spot	Unidentified disease	3	'The plant turns black, completely black. We cure it (spray fungicides) but sometimes we can't because there is so much disease.'
Ceniza	Ash	Mildew?	2	'Sometimes when it is cold the whole plant turns black.'
Pudrición seca	Dry rot	Storage disease	1	
La negra	The black one	*Erwinia* sp.?	1	

Table 4.6   Arthropod pests, according to farmers

Folk name	Meaning	Identification	No. of times mentioned	Observations by farmers
Palomilla Polilla Mosquilla Candelilla Mariposa	Moth Small fly Small candle Butterfly	Lep: Gelechiidae	6	'The little candle is a small butterfly. Moth, little candle, it's the same thing. It occurs in potatoes in storage. One cannot eat potatoes if they have moth in them. They become bitter. Not even the pigs will eat them.' 'We do not store potato, because a worm, the little candles, makes holes in them. We can store potatoes for a maximum of two months. Nobody stores seed.'
Saykuro Gusano del choclo, Choclo kuro, Taykuro, Tushkuro	Most of these names are (or probably are) variations on the name 'maize worm.'	Larva of *Spodoptera* spp. (Lep: Noctuidae)	6	'It cuts off the stem; it comes from out of the earth.' 'It is big. It makes holes, like little wells. It cuts off the plant when it is coming up. And it makes holes in the tubers.'
Pulguilla, Pulgón, Mosquita, Shipe, Shipo	Little flea Big flea Little fly (last 2 from Quechua?)	Adults of *Epitrix* spp. (Coleoptera: Chrysomelidae)	6	'It is a small worm. It makes holes in the potato. It is black and small and it hops about on top of the leaf.' 'It is black, like a beetle (escarabajo). It is on the leaf.'

Lorito, Loro, Loro verde, Gusano verde	Little parrot Parrot Green parrot Green bug	*Diabrotica* spp. (Coleoptera: Chrysomelidae)	5	'It eats the leaf. It is a little worm and it makes holes in the leaf and so we spray it even more.' 'It is really small. It has dots on it. It is green and it flies.'
Vaquita	Little cow	Probably adults of several spp. of Coleoptera, e.g. of the Coccinelidae family.	3	'Vaquita is a little fly. When it smells the medicine (pesticides) it leaves. There are blue, green, yellow and red ones. There are different kinds.'
Gusano blanco	White worm	Larva of the Andean potato weevil.	3	'A worm that eats the plant. It is white and it makes the potatoes maggoty in the earth.'
Mosco, Mosquillo	Fly Little fly	Possibly a small Homoptera, e.g. *Empoasca* spp.	2	'It is very small. It gets together in groups, inside the plant.'
La chía	Probably from the Quechua *ch'iya*, meaning 'nit.'	*Meloidogyne* sp. nematodes.	2	'It attacks the fruit, which becomes white.'

Comparing the above tables, the correspondence between the técnicos' report of local names, and the farmers' statements is consistent, but not identical by any means. Among others, farmers gave us more synonyms for most diseases than the técnicos did, but técnico's reported more precise definitions. The técnicos gave several explanations that would have been difficult to tease apart just through field interviews e.g.:

- *Virosis* is used to describe 2 viral diseases, but that there are other synonyms for each disease.
- Marchitez labels the bacterial wilt disease, but 'la pus' describes one of its most noticeable symptoms.

Besides describing their knowledge of crop damage (Tables 4.5 and 4.6), farmers also told us about disease aetiology and control measures, which we have not listed here. (See Bentley & Vilca 2001a, 2001b.)

Tables 4.7 and 4.8 compare the overlap between names mentioned by farmers and técnicos for diseases and pests. The most important crop health problems were reported by both groups of people, but quite a few were only mentioned by one or the other. A rapid survey that did not consult both groups would probably miss a large percentage of the total named categories of pests and diseases. Técnicos and farmers probably mentioned slightly different categories because:

- Some pests and diseases are found in only a few places, and the técnicos are familiar with a larger area than most farmers.
- Agricultural research tends to replicate experiments over several years, which helps to maintain long-term interest in certain organisms, while farmers are more interested in the pests that have been most important in the past year or 2.
- Farmers are concerned about some organisms, especially insects, which are conspicuous but not especially damaging. Farmers occasionally over-estimate the damage of e.g. *Diabrotica* spp. in the Americas (Bentley & Rodríguez 2001), or the stem borer in rice in the Philippines (Heong & Escalada 1999).

Generally, the técnicos' descriptions of folk categories were accurate. But there were a few mistakes. For example, the técnicos said that 'vaquita' referred to the small green leaf beetles (*Diabrotica* spp.), but farmers called the leaf beetles *lorito* (or a similar word) and used the word vaquita as a residual label for many kinds of small beetles (including lady bird beetles and small scarabs). In other words, the técnicos' accounts of local categories were an excellent start, but still needed to be verified and complimented with farmer interviews.

Table 4.7   Comparison of folk names of diseases, as given by técnicos and by farmers

Folk name as given by técnicos	Folk name as given by farmers
Rancha	Rancha, Rancha negra, Hielo fungoso, Quemazón, La plaga
Marchitez	Marchitez
Bacteria	–
Cáncer	–
La pus	–
Liendrecillo	La chía
Pudrición	–

Seca seca	–
–	Hielo,
–	Granizada
–	Mancha negra
–	Ceniza
–	Pudrición seca
Ticte	–
Virosis Plantas cansadas	–
Plantas amarillas VirosisHoja de lima	Amarillas Virosis

Table 4.8   Comparison of folk names of pests, as given by técnicos and by farmers

Folk name as given by técnicos	Folk name as given by farmers
Rancha amarilla	–
Candelilla Polilla	Palomilla Polilla Mosquilla Candelilla Mariposa
Gusano de tierra Huyo	Saykuro Gusano del choclo, Choclo kuro, Taykuro, Tushkuro
Vaquita	Lorito, Loro, Loro verde, Gusano verde
–	Vaquita
Mosquilla Shipe	Pulguilla, Pulgón, Mosquita, Shipe, Shipo
Zorrillo	–

–	Mosco, Mosquillo
Chamson	–
Gusanera	–
Gusano blanco Gusano arroz	Gusano blanco

New knowledge, better development

We hope that this brief introduction will stimulate others to explore a new way of using a neglected source of knowledge. Our main point is that local technical people are experts in local conditions and local rhetoric. Técnicos (especially extension agents) are not only specialists who broker information from scientists to farmers, but professionals who can explain (many) folk concepts in scientific terms.

Notes

1   To paraphrase V.S. Naipaul, life in rural India improved a great deal in the 30 years after 1960, because of the people who went about development in the best way, by doing it as their job (Naipaul 1990).
2   Scientific names are also derived from Greek and other languages.
3   From the Quechua 'qhoña' meaning snot.
4   Unlike folk names, which vary from place to place (they have high geographical synonymy) and scientific names, which change through time with taxonomists' publications.
5   *Pilpintu* is the Quechua word for moth/butterfly. Its use here reflects a mistaken association of lepidoptera with this maggot.
6   Note that many of the same words are used to refer to aphids and to peach leaf curl. Some farmers confuse the two.
7   This column was added later by the anthropologist.
8   Luis Quispe, agronomist with the Peruvian crop and livestock health agency (SENASA) participated in the fieldwork and helped identify several of the diseases.

References

Bentley, J. W. & G. Rodríguez. 2001. *Honduran folk entomology. Current Anthropology* 42(2): 285–301.
Bentley, J. & P. Vilca. 2001a. *La papa en Huánuco: Semilla y conocimiento popular sobre las plagas y enfermedades.* Lima: Report submitted to CIP.
Bentley, J. & P. Vilca. 2001b. *La papa en Huánuco: Semilla y conocimiento popular sobre las plagas y enfermedades.* Lima: Report submitted to CIP.
Berlin, B. 1992. *Ethnobiological classification: Principles of categorization of plants and animals in traditional societies.* Princeton: Princeton University Press.
Chambers, R. 1980. The small farmer is a professional. *Ceres* 13(2): 19–23.

Ellen, R. F. 1993. *The cultural relations of classification: An analysis of Nuaulu animal categories from Central Seram.* Cambridge: Cambridge University Press.

Heong, K.L. & M.M. Escalada 1999. 'Quantifying Rice Farmers' Pest Management Decisions: Beliefs and Subjective Norms in Stem Borer Control.' *Crop Protection* 18: 315–322.

Hunn, E. 1982. The utilitarian factor in folk biological classification. *American Anthropologist* 84: 830–847.

Naipaul, V. S. 1990. *India: A million mutinies now.* London: Minerva.

Padilla A. M. 1999. *El Cultivo del Duraznero en los Valles Cruceños.* Santa Cruz, Bolivia: CIAT.

Van Mele, P. 2000. *Evaluating farmers' knowledge, perceptions and practices: A case study of pest management by fruit farmers in the Mekong Delta, Vietnam.* Doctoral thesis: Wageningen University, the Netherlands.

Chapter 5

Local History as 'Indigenous Knowledge': Aeroplanes, Conservation and Development in Haia and Maimafu, Papua New Guinea

David Ellis and Paige West[1]

In this chapter, we demonstrate the importance of local history, in both theorised conceptions of indigenous knowledge ('IK') and in the practice of development and conservation in Papua New Guinea. We place our emphasis on a broad definition of local history in relation to debates on indigenous knowledge and specifically on 'local knowledge'. We present an evaluation of the connotations of the term 'indigenous knowledge', critiquing the predominance of scientific value and biological and economic prerogatives often implied by its use. In this respect, we argue for a shift in focus, in both theory and practice, towards an engagement with history in considerations of local knowledge. This would signal a move away from notions that are restricted to biological or economic relations between people and the 'environment' in favour of broader, more anthropological concerns.

Historically, people of the two distinct ethnic groups with which we have worked in Papua New Guinea[2] have had relationships based on the exchange of women and sago and on alliances during times of regional conflict. The boundaries of colonial administration and then national government have been drawn across the mountains between these two groups. Since the 1980s, they have experienced a further set of boundaries within which their lives and relations are enacted – those of a single 'conservation and development' initiative.[3] The findings of our research indicate that local histories have been both ignored and rewritten within the context of conservation and development.

The paper draws on ethnography and narratives about grass airstrips and aeroplanes in the history of development on the lands of both groups. We suggest that local history is far more diverse than biologically or economically oriented representations of people, 'culture' and 'knowledge' would suggest.

Defining 'local history'

Sillitoe (1998) employs the term 'local knowledge' in preference to other labels in the elaborate jargon of indigenous knowledge and development.[4] We situate the starting point of our argument broadly within his definition:

Local knowledge in development contexts may relate to any knowledge held collectively by a population, informing interpretation of the world. It may encompass any domain in development, particularly that pertaining to natural resource management ... It is conditioned by socio-cultural tradition, being culturally relative understanding inculcated into individuals from birth, structuring how they interface with their environments (ibid.: 204).

Concordantly, Posey (1998) defines 'Traditional Ecological Knowledge' (TEK) in the following way:

TEK is far more than the simple compilation of facts ... It is the basis for local level decision-making in areas of contemporary life, including natural resource management, nutrition, food preparation, health, education, and community and social organisation. ... TEK is holistic, inherently dynamic, constantly evolving through experimentation and innovation, fresh insight and external stimuli. ...One area where TEK is well understood and exploited is that of agriculture (ibid.: 96–7).

The two definitions make implicit reference to history – Sillitoe mentions tradition and the trajectory of the life of an individual from birth, and Posey writes of a dynamic and evolving process. We wish to take this implied interest in change in knowledge over time a step further on three counts – with respect to history, biology and economics.

Firstly, we propose a definition of 'local history'.[5] Our conception of local history arises from a range of narratives, concerns and lived experiences of the people with whom we worked in New Guinea. It incorporates what might be termed the expression of 'culture', or, in Papua New Guinean English, 'custom': principles of social grouping and organisation, kinship, marriage, exchange, ritual life, oral narrative, mythology and song. Within this, we think of human ecological history as an integral part of social life:[6] subsistence activities, such as gardening, hunting, fishing, tending fruiting trees and palms and harvesting their yields, and gathering foods and materials from the forest. We are also concerned with how these activities have changed over time, how relations between people and land are conceptualised and how they themselves have changed. Our vision of local history is of a living, changing process, at the levels of structure, practice, belief and expression. It is often enacted in complex forms in particular events. It differs from perspectives offered by IK research in that it is not framed first and foremost in terms of how people interact with the environment. Instead, it is focused on experience, narrative and performance, striving to incorporate the metaphorical, metaphysical and practical ways in which these are enacted and articulated in people's lives. This focus on oral narrative and articulation is marked out not only in what people say to each other and how they interact (and not only in the presence of the ethnographer) but also at the levels of ethnographic method and writing. In other words, an ethnography of local history must give space and time for narratives to be listened to, heard, recorded and written into ethnographic texts. Local history can be oral, written or just experienced; it is not dependent on writing or bureaucratic procedure. It also includes the articulations of people's desires for change and development and their own conceptions of distant and recent change. It is not restricted in terms of time or space. It incorporates the perspectives of women, men and children. It is not necessarily collective; it may be individualised or intertwined with other histories.[7]

Our second emphasis concerns both biology and ethnobiology, particularly in theoretical formulations of indigenous knowledge. Even more recent, inclusive and 'culturally' informed definitions, such as those presented above, tend to focus on 'biological' relations between people and environment.[8]

The third strand of our argument focuses on the impact of free market economics on conceptions of local knowledge. Local knowledge has become a commodity in development, partly through a disproportionate emphasis on the value of biological knowledge. We have found that biological and economic prerogatives are embedded deeply within representations of indigenous knowledge.[9] This would suggest that the demands of research and development programmes have a significant influence on local history and local knowledge.

Evaluating 'indigenous knowledge': culture as biological 'information'

Universal applications of the language of 'indigenous knowledge' are contentious. In Papua New Guinea, for example, most people living in rural areas are the customary inhabitants of the land,[10] with the exception of a minority of expatriates. In such cases, the concept does not have particular currency in terms of making a distinction between autochthonous inhabitants and descendents of settlers or expatriates.[11]

Indigenous knowledge, described in broad terms by Warren, Slikkerveer and Brokensha as, 'the local knowledge that is unique to a given culture or society' (1995: xv), is often represented in over-specialised ways. Scholars of indigenous knowledge tend to focus on people's relations with the natural world, either through agriculture (see Conklin 1954; Richards 1985; Nazarea 1998), subsistence in general (Fowler 1977), the conservation of biodiversity (Gadgil 1993), or how the local biological world is conceptualised and classified (Berlin, Breedlove and Raven 1973; Bulmer 1976). In other words, indigenous knowledge is often recognised as knowledge of the plant or animal world before it is recognised as sociality and history or as a social and historical product. Taking this a step further, it is often seen as a biological referent, a 'folk' knowledge that might be of interest to biological research, but is nevertheless inferior to 'scientific' data.

Representations such as these are not unique to natural science. They are often implicit in works of ethnobotany, human ecology and anthropology. Although an academic focus on plant knowledge is not intended to set up divisions between the environmental and the social or historical, we suggest that this can often be the end result.[12] While it is difficult to interpret biological or ethnobiological data without an understanding of local sociality and history, some commentators on the relations between people and the environment continue to deprioritise or overlook this. A focus on biological prerogatives in local knowledge has important implications for power relations in general. There is a risk for local people to be perceived merely as conduits for research or development, and not as equal, social human beings.[13]

Anthropologists and other social and natural scientists that have written about indigenous knowledge often make reference to the notion of 'value'. In accordance with our aim here to make an evaluation of the category 'indigenous knowledge', we turn to discussions of value – values placed on indigenous knowledge, values of scientists and values in biology.

We have already identified a biological emphasis within representations of indigenous knowledge. Although there has been increasing recognition that indigenous knowledge as cultural heritage should also be valued (Posey 1998),[14] such perspectives also suggest that the driving force is free market economics:

> Although international efforts to recognise indigenous, traditional and local communities are welcome and positive, they are pitted against enormous economic and market forces that propel globalisation of trade (ibid.: 100).

Friedberg (1999) and Shiva (1993) share this observation:

> ...the laws aiming to preserve biodiversity, elaborated at great pains in the international conferences that took place after Rio, hardly hide the fact that the market is still the ultimate value in our societies (Friedberg 1990: 2).

> ...while biological resources have social, ethical, cultural and economic values, it is the economic values that must be demonstrated to compete for the attention of government decision-makers (Shiva 1993: 170).

Posey (1998) argues that one of the major shortcomings of globalized economics is that: 'value is imputed to information and resources only when they enter external markets,' (ibid.: 100). Shiva's analysis complements this:

> Nature's diversity is seen as not intrinsically valuable in itself, its value is conferred only through economic exploitation for commercial gain. This criterion of commercial value thus reduces diversity to a problem, a deficiency (Shiva 1993: 164).

This presents us, in turn, with a difficult problem. If value can only be achieved in a global market economy through the entrance into that market, where does this leave local biological or environmental knowledge and, more pertinently, indigenous knowledges? The obvious answer is that indigenous knowledge becomes a kind of 'information' in a market economy and is thereby commoditised. It is partly in response to this that we emphasise the importance of local history. 'History' and social life have worth in their own right, just as biodiversity has, 'intrinsic value', as Shiva puts it (1993: 172). The pursuit of short-term market-oriented goals does not lend itself to engagements with complex and contested fields such as history and social life, however.[15] In response to this, we argue that when 'biological knowledge' is disembodied from its cultural and historical context, its value is diminished.

Yet the pervasion of language and concepts of free market economics can even be noted in writings that set out to emphasise the importance of sociality in these debates. Posey, for example, asserts that, 'Industry and business discovered many years ago that indigenous knowledge means money' (1990: 14). He proceeds to present a discussion of the 'annual world market value for medicines discovered through indigenous plant knowledge (ibid.: 15). He focuses on economic value as a harsh reality, hence his main argument for the need for adequate compensation mechanisms to support indigenous people and their knowledge.[16] He proclaims:

> If something is not done now, mining of the riches of indigenous knowledge will become the latest – and ultimate – neo-colonial form of exploitation of native peoples (ibid.: 15).

This is a subtle case of the language of free market economics being used to describe indigenous knowledge, 'mining its riches' being the image in question. In a 1998 paper, Posey contrasts indigenous knowledge with free market economics: 'sacred balance' as opposed to the 'balance sheet'. Yet a language of economics can serve to overshadow the value of local knowledge, even in a paper that privileges social perspectives.

Nazarea (1998) presents a justification for the concept and method of 'memory banking'. She, in turn, paints the same gloom-laden picture as evoked by other commentators about the overriding force of economics over knowledge – of, 'a production and distribution system that relies on streamlining and simplifying agriculture for greater efficiency and profit' (ibid.: 4). Knowledge of the diversity of the plant world in agricultural practice is, to some extent, perceived as 'obsolete' (ibid.: 5) in such a scenario. With regard to the preservation of plant and genetic materials in herbaria and gene banks, a situation could arise where the biological and genetic information stored would be knowledge that had expired if its social context in human memory had been lost. Through the employment of social scientific methods to research the history of local agricultural knowledge, Nazarea advocates the establishment of 'memory banks' to supplement genetic and other 'banks', effectively putting such 'cultural' knowledge on hold, in a frozen state, until such time as it might inspire use value or interest again. She writes:

> ...certain parallels can be drawn between a gene bank and a memory bank. While germplasm encodes genetic information that has evolved through time as a response to selection pressures, cultural data in the minds of local farmers who have had considerable experience in growing these crops are repositories of coded, time-tested adaptations to the environment (ibid.: 6).

This is a pertinent example of perceiving indigenous knowledge on the terms of biological (and genetic) information. Of course, we recognise the importance of attempts to document local history, including that of ecological change. However, if we evaluate the language and conceptual terms in which they are framed, the style often appears to embody the forces that erode and negate local knowledge. Although Nazarea bemoans what she refers to as, 'mindless homogenization' (ibid. 14), in some respects her treatise might be more pertinent to the biotechnology industry than to people's lived experiences of local history. This also begs the question: to what extent can ecological knowledge or local history be disembodied from social experience? Who has the right to employ, interpret or transform 'banked' knowledge? (Strathern 1999).

We have perhaps to ask ourselves: What could be the effects of applying the language and concepts of the free market economy to indigenous knowledge? From the perspective of women in less consumerist regions of the world, Shiva offers a valuable insight:

> There are a number of crucial ways in which the Third World women's relationship to biodiversity differs from corporate men's relationship to biodiversity. Women produce through biodiversity, whereas corporate scientists produce through uniformity. For women farmers, biodiversity has intrinsic value – for global seed and agribusiness corporations, biodiversity derives its value only as 'raw material' for the biotechnology industry. For women farmers the essence of the seed is the continuity of life. For multinational corporations, the value of the seed lies in the discontinuity of its life. Seed corporations deliberately breed seeds that cannot give rise to future generations so that farmers are transformed from seed custodians into seed consumers (1993:172).

In spite of our critique of what we see as the bio-centric focus of indigenous knowledge research, we acknowledge that IK-oriented research has the potential to value local perspectives. In contrast, in some conservation and development projects local people and their cultural practices and histories are seen as impediments to the goals of the projects. For indigenous knowledge research to influence conservation and development projects for the benefit of both local people and project outcomes, we argue that researchers must embrace local history and experience as an integral part of their models of 'knowledge'. Below, we will argue that for conservation and development interventions to be successful, practitioners must value local history and social practices. In the following section, we discuss depictions of local people by conservation and development practitioners associated with the Crater Mountain Wildlife Management Area in Papua New Guinea.

Depicting local people

In the 1990s, the public role of anthropology in representing local people in development was often reaffirmed (Pigg 1992, 1993; Hobart 1993), sometimes with regard to indigenous knowledge research (Warren et al. 1995; Sillitoe 1998a, 1998b). Chambers and Richards (1995) write of the influence anthropology has exercised in representations of indigenous knowledge since the 1970s. Yet they are cautious to highlight the complexity of the relationship between social science and development:

> The reality ... has changed less than the rhetoric. The awareness, attitudes and behaviour of many development practitioners have changed less than the language they have learnt to use. Many have acquired the easy skill of using words like 'participation' and even 'empowerment' but without changing the way they see poor people or the way they feel development should be undertaken. The language has become bottom-up but the inclination remains top-down. ... [T]he great majority of development professionals undervalue indigenous knowledge and the capacities of local, especially rural, people. For this majority of professionals, 'they' and 'what they do not know' are still the problem; and 'we' and 'what we know' are still the solution (ibid.: xiii).

In our work with people in Papua New Guinea,[17] we have observed perceptions such as these played out on a daily basis in the context of a conservation and development project. In spite of the language of 'community based conservation' (Western et al. 1994) and 'participation', we have encountered persistent depictions of local people and their lifestyles and customary practices as a 'threat' to biodiversity (James 1995: 5; Johnson 1997: 396), as 'constraints' and even as an 'inherent problem' (Johnson 1997: 420). Even if the significance of history and social life is acknowledged, the tendency is to suggest quite openly that local people must change to meet the demands of externally generated development, and not the other way round.[18] People's customary lands are also presented in a pejorative light: their 'remoteness' and what they 'lack' are emphasised (Johnson 1997: 396).

Representations of how people interact with their land and its 'biodiversity' have not been based on data on human ecology in the Crater Mountain project. References to the impact of human population on local ecosystems have often been speculative, based on generalisations and scant anecdotal evidence[19] (RCF and WCS 1995: 12). They are also motivated by Euro-American conceptions of those ecosystems. For example, there is no assessment of the relative value of rat in the diet of children or in local perceptions of this as a food; it is merely used strategically to argue that populations of larger game animals have been depleted:

> Most grown men remember fondly early hunting trips with their fathers when they would return to the village from the forest with plenty of game. Today, when one sees village children eating meat, it is most likely wild rat (RCF and WCS 1995: 12).[20]

The discursive production of local people and their actions as a 'problem' to be endured by conservation practitioners has, in itself, problematic outcomes. It often translates into situations where conservation organisation employees experience village events where important local knowledge is enacted or acted upon and, instead of trying to understand these events, they articulate disgust with local people. The publication of repeated representations of people as a hindrance to the smooth running of conservation projects lends a self-referential justification to these representations. This process negates engaged analysis or recognition of the complexity of situations in which conservation practitioners, as well as local people, are fully implicated.

Depictions of the lives of people living in highly biologically diverse rural areas, in both conservation and development publications and in the daily speech of conservation and development practitioners, are not simply linguistic representations. They are politically motivated discourses that help to shape the future trajectory of local development and people's influence within that (Brosius 1999). The power differential in conservation and development projects is weighted heavily towards project practitioners, not towards local people. When biologists, development workers, volunteers and others perceive and depict the social milieu of rural landholders in ways that do not reflect local lived experiences and histories, the images and ideas created and maintained affect conservation and development practices and outcomes and, ultimately, local people.

Through the power of the written and spoken word in development discourse, the perceived failures of the transnationally funded Crater Mountain project are being attributed to local people. While certain conservation biologists have suggested that project failures have been caused by local social constraints, we make an alternative argument. We propose that local history be considered before any development or conservation project is initiated, and that anthropological research should become a basic part of the planning of such programmes. We believe that a methodology which values social and historical experience through an engagement with anthropological research would move beyond antagonistic representations which seek to explain the causes of development failures in politicised terms.

In the following two sections, we have chosen to focus on local narratives about aeroplanes and airstrips to illustrate the importance of historical perspectives in approaching an understanding of development and change. In conservation discourses in Papua New Guinea in general, local people are often referred to as being 'cargoistic' in their reactions to development[21] (Grant 1996: 7–8; McCallum and Sekhran 1997: 53, 54–5; Ellis, J-A. 1997: ix). We have also encountered such representations in interactions with conservation biologists working on the Crater Mountain project (see Bickford n.d.). In this respect, the reaction of local people to aeroplane landings is depicted as being one of 'cargoism' or greed. Conservation biologists and others working in the region invariably have to fly in, along with research equipment and supplies, by aeroplane or helicopter, to the sites where they work. Quite apart from the question of correlations between aeroplanes laden with researchers, equipment and provisions and representations of local people's reactions to researchers and their 'cargo', we demonstrate below that local conceptualisations and experiences of aeroplanes are an important part of local history and knowledge. In a sense, we present them as 'indigenous knowledge' in their own right to demonstrate the importance of all kinds of historical narrative in local experience.

In the Pio-Tura region and in Maimafu village aeroplanes and aeroplane landing strips are a conduit for conservation, development, and commodities. They are also much more than that. They are the social and historical sites of local experiences of development and their ethnography is an important part of our individual research projects on local history.

Coffee, religion and pigs: the airstrip at 'Maimafu'

There are no pigs in Biabitai, Motai, Tulai, Abigarima, Kolatai, Harontai, Lasoabei, Kuseri, Atobatai, Bayabei, Iyahaetai, Wayoarabirai, Kalopayahaetai, Halabaebitai, or Aeyahaepi. These ridge top settlements make up what has become known, through government and conservation organisation discourses, as Maimafu Village.[22] A 'village' in the highlands of New Guinea with no pigs was, in the past, an unusual occurrence. Pig husbandry is often classified as 'extensive' there (Morren 1986: 88). Among Gimi speaking peoples, including the people in Maimafu, pigs have traditionally been an extremely important part of subsistence, brideprice, ritual, and other social practices (Gillison 1993). Today, a number of places in the Highlands have eradicated pigs and have ceased historic practices associated with pig husbandry. Maimafu is one of these places.

Maimafu is located culturally on the border between the 'central highlands' and the 'highlands fringe' culture areas, and land held traditionally by residents of Maimafu spans a topographic range from lowland rainforest to montane cloud forest. People there practise shifting cultivation to produce a variety of foods and the area is known throughout the region for its abundance and quality of garden produce. Some important staple foods are sweet potatoes and other tubers, green vegetables, corn, pumpkins, and peanuts. People also collect what they term 'wild' foods from the forest. Women and men spend much of their time tending coffee gardens, as the sale of coffee provides the most substantial portion of cash income. Most people living in the settlements that make up Maimafu claim membership in the Seventh Day Adventist (SDA) church. Among the residents of Maimafu, events defined locally as 'development' include the opening of the village aid post in 1984, the opening of the 'community school' in 1991, the completion of the airstrip in 1992, and recent prospecting for gold by Macmin N.L., an Australian mining company (West 2000: 94–190).

The absence of pigs in Maimafu is tied to both religious prohibitions and pragmatic individual choices. In 1982 the adult men there 'voted' to kill all their pigs and end local pig husbandry 'forever'. At the same time they 'voted' to build, with the help of the Seventh Day Adventist Church, an airstrip.[23] There are various stories about why people decided to build the airstrip, but the one told most often is about carrying coffee to the 'head of the road' in a distant village:

> In 1980 my wife and I carried seventy kilograms of coffee and our baby to Ubaigubi. We wanted to sell the coffee in Goroka, so we had to carry it to a place with a road. She carried fifty kilograms, and I carried twenty and the baby. It was a terrible walk. We went up over the mountains, and I hurt my back as we crossed. I sat down in the road and cried. I told my wife that if I made it to Ubaigubi alive, I would stand up for government election and build a road or an airstrip to Maimafu. I cried because the government forgot my people. I cried because they see us as useless, and they do not help us. I promised my wife that if I won, I would build an airstrip, build a road, start a school, and build a hospital.

Kayaguna Kelego recounted this version of the coffee-carrying story. Kayaguna is approximately forty-seven years old and is one of the most well respected men in Maimafu. In 1966, upon the urging of a locally based SDA missionary, Kayaguna was the first man from Maimafu to go out of the village for schooling. After graduating from the mission school he worked from 1970 to 1974 as a Seventh Day Adventist missionary. He returned to Maimafu in 1974 and subsequently became a local elected official.

The building of the airstrip is one of the most salient events in the history of development in Maimafu. Both men and women mark its building as the beginning of 'getting services', and as the end of pig husbandry. Pigs were a way of life prior to the construction of the airstrip. Although some people cite adherence to SDA food prohibitions as the primary impetus for this change in subsistence practices, the negative effects that pigs would have on the airstrip were a primary factor in this decision. Although the construction of the village aid post began before the construction of the airstrip, and the aid post was completed before the first plane landed in Maimafu, people see the airstrip as their link to 'health', 'development', and the rest of the country. Kayaguna says,

The first plane came in 1992. It was a SDA plane. The church had given us tools to build the airstrip. They brought them in on a helicopter and we started working. They were the first to use the airstrip. When the plane landed we sat down and cried. That airstrip is our link to development. All of what we have comes through it. People here believed me in 1982 when I sat down with the men and told them my idea of the airstrip. They started to build it. It took ten years to finish it. They worked hard. Not just men, but the whole village. We worked to make this happen, and it did.

The people here built it for nothing. They did not get paid for the first nine years of work. When I was elected in 1991, I got some money from the national government, but it was very little. I got the people one year's wages for ten years of work. Like I said, it is our tie to the world. People can sell their coffee and get money now. People can go to Goroka if they are very sick. They can get things from stores now. They can get things that they need.

In many ways the airstrip is a major focus of life in Maimafu. Trade store goods, visitors, and information enter Maimafu via the airstrip, and coffee, the only local cash crop, exits the area via the airstrip. An aeroplane landing is a big event. When an aeroplane is scheduled to land, people from all of the ridge tops go to the airstrip to wait and waiting for the plane has become an important social event in Maimafu. People use the time to visit with relatives who live on different ridge tops, discuss current social events including weddings, the creation of new gardens, instances of sorcery and the like, and talk about current coffee prices. People also use the time to talk about what may or may not be coming on the plane and where and who it came from. The point cannot be made too strongly – everything that comes to Maimafu from 'elsewhere' comes on an aeroplane.

The connection between these aeroplane landings, the airstrip, and the rest of the world can be seen in people's discussions about the pilots. Everyone in the village knows every pilot's name, where they are from, what their family make up is, what church they go to, whether they like peanuts or not, and a whole myriad of other personal details. In addition, people know what aeroplane is going to land (including the make and model) and who the pilot is by the sound of the approaching aeroplane. 'Local knowledge' about aeroplanes is extensive and complex and it is fundamentally tied to local development desires. This 'local knowledge' is also tied to the social lives of people in the village in that some of the pilots who land in Maimafu are SDA missionaries. These men become part of the socio-religious landscape for villagers as they make social connections with people, frequently fly in for and attend Saturday church services in Maimafu, and make friendship bonds with people.

Although MAF and SDAA are the only planes that fly regular routes that include Maimafu, other planes occasionally land there. Upon an unexpected landing there is a general feeling of excitement, elation, apprehension, and interest. The fervour generated by these unexpected flights goes on for weeks afterwards. For instance, one sunny afternoon a plane belonging to an evangelical American mission group (which works in the region but not in Maimafu) landed in the village. When two American men, dressed identically in the style of the American movie Top Gun with Levi's jeans, brown leather flight jackets, crew cuts, and mirrored sunglasses, left the plane there was an uproar. It was quite a sight. The two men, upon seeing a white woman, immediately confronted her and did not speak with the older men who are considered local leaders. They spoke in English and, without allowing her to ask any questions, began loudly and familiarly throwing questions at her. After finding out where they were – they were looking for a different village – they both, without shaking anyone's hand or saying 'thank you', got back in the plane and left. This 'incident' as it became known in the village was discussed for the next month. It was termed a 'big something' and an 'affront' or 'insult' to the community. There were late night discussions about writing a letter to complain about the 'rudeness' of their pilots since they had not spoken to 'even one man' from the village. There were lengthy question and answer sessions about what exactly had been said. And, finally, there were numerous conversations about the sunglasses. These were the first mirrored sunglasses seen in Maimafu.

The above ethnographic moment is recounted to make the point that the airstrip is now where change and difference enter the community. It is the site of new things, ideas, people, and money. It is also an historic source of change: pig husbandry no longer takes place in Maimafu because pigs are thought to damage airstrips. SDA missionaries, who as mentioned above, helped with the airstrip construction, 'warned' the residents of Maimafu that pigs 'bring trouble' to airstrips. All of the Crater Mountain project activities are centred close to the airstrip, and all of the conservation and development organisation affiliates who come into the village enter that way. All of the teachers, although they rarely stay a full term in the village, come on planes and live next to the airstrip. The community school is immediately next to the airstrip. Everything associated with 'development' is also associated with aeroplanes and the airstrip.

A strong man is coming: aeroplanes and the Pio-Tura region

Haia village lies about 30 kilometres south of Maimafu. Although this is not far as the crow flies, the two villages are separated by an almost unsurpassable range of pinnacles referred to in English as 'Crater Mountain'. Very few people undertake the treacherous journey on foot. The landscape in Haia is markedly different to that in Maimafu. While Maimafu lies at over 2000 metres above sea level, Haia is situated at an altitude of less than 800 metres. The terrain is still rugged but Haia is surrounded by dense forest as opposed to mountain ridge tops. The lands of the Pio-Tura region around Haia are inhabited by people often referred to as 'Pawaia' who speak a language considered by them and others to be a linguistic isolate and who have a distinct set of customary practices. Sago is the staple food there, although people also plant swidden gardens and tend areas of fruiting trees and palms. They also gather, hunt and fish in the forest. Within a topography of 'culture areas' in New Guinea, Haia might be located around both the 'highlands fringe' and 'Mountain Papuan' areas in the region between the highlands and the southern coast with influences and relations on both sides (see Weiner 1988).

A grass airstrip was completed in Haia in 1975 by manual labour. This was a result of the establishment of a New Tribes mission. The expatriate missionaries who arrived in Haia in the early 1970s were still there in the early years of the new millennium. While Christians in Haia feel an allegiance to New Tribes in terms of spiritual teaching, people also connect the subsequent changes in the region to the mission. Health care was provided through the mission and then, from 1989 onwards, through a government health sub-centre. A primary school was established in Haia in 1991. More recent incursions into the region have included representatives from logging companies, conservation biologists and ecotourists. Cash crops have been less successful in Haia than in Maimafu. Very little coffee and other cash crops are produced here.

Although the airstrip in Haia has a distinct history, there are many parallels between people's perceptions of the airstrip in Haia and in Maimafu. One day in September 1999, an old woman called Nira, who was sick and walks only with a stick, was asked: 'When you first saw a plane, how did you feel?' She replied by describing a plane landing and her own and others' reactions to it:

> When I see him coming, I say: A strong man is coming now. I say this: This strong man, the man who goes around the clouds, he is coming now. You young people, go quickly! Your boy is going now – you must go quickly! He will not stay long, he will go back. He is coming just to show us and he is coming back now. I sit on the veranda of the house and I sing like this. When I sing like this, the young boys laugh at me. The pilot sits down at ease in the plane and he drives the plane and he is flying now. He smiles and pilots the plane and he flies. He looks as though he is really smiling and he is flying, and later on, the people who wear long clothes, it is your time to be happy and make these long clothes of yours dirty. I sing like this and laugh as well as I sit at my house. When the pilot flies away, he looks at me with a smile. The plane flies in the clouds, and he flies away. You people in the houses, go quickly, the plane is going to land now. And all men and women gather people together and set off. All of them go to see the plane, and I will sit down alone at my house and I will sing. When the men and women all set off in hoards to see the plane, and the plane gets back up and goes like that, I will see. I will turn my face up to the plane which is coming in, and I will watch. Yes, I will watch, and when he turns his hands like this [the wing and tails of the plane will wave and move before flying], I know that he is coming. When he turns his hands, I know that he wants to return now. The plane goes around in the clouds and the clouds stick to his propellers.[24] He wants to go and he is starting to move now. I will say this: You women – move quickly! All the men, all of them will move and go, and I will stay at home on my own. I will just sing at the house. When the plane flies away, the men will run after this plane. People call my name and they say, 'Ai, mother, you must come!'. And I come back at them by saying, 'It would not be good if I fell'. People tell me to come and I say: I cannot come. He will not stay here for a long time. When I see him, he will fly and go up and away again. This is how I sing: He will leave me here, and he will go up and away again. This is how I sing about it.

Nira employs metaphor in her narrative to bring the event to life retrospectively, and also to rework it creatively as it happens. She uses song as a vehicle for this. The aeroplane and pilot combine, almost as one being, the 'strong man'. The plane is somehow an extension of the pilot, a single being. It is not possible to distinguish who is turning the hands, who the strong man is, the pilot or the plane. The aeroplane landing becomes her performance. She describes the structure of any aeroplane landing in Haia village. People drop what they are doing and head for the airstrip if they can. Nira is unable to make it but she shows us how important an event an aeroplane landing can be. The first sounds of the aeroplane as it approaches and cuts through the other sounds of the day are striking.[25] People on the other side of the valley who are unable to reach the airstrip quickly also drop what they are doing instantaneously. It is everyone's event.

Nira's reaction is one of joy and her singing is almost ceremonial. Her descriptions of the smiling pilot evoke a person with whom she has the potential for relationship, even though they have never met. She is full of laughter, and she makes fun of all the people who rush to the scene of the landing of the plane. She mocks their long second hand clothes, symbols of technological change alongside the ultimate marker of change, the aeroplane. So, an old woman who cannot walk to the airstrip brings this event to life in a lively narrative full of metaphor, satire, song, performance and laughter.

In the late 1990s, Tipe, an old man in his mid-sixties, described the first sighting of aeroplanes flying over his region as follows:

I showed my father what I saw when the Japanese were starting the war with the government. I saw an aeroplane. We were cutting sago when we saw this plane. We were about to take a stone axe to cut down the sago palm, and we took out the heart of the sago palm, and I ate the heart of the palm with sago. ...

We used to cut sago, and my uncle cut sago. His name was Epiae. My uncle, Epiae, cut it. My uncle who was called Anapi removed the outer bark of the sago with a stone knife. My uncle Jonli removed the heart of the sago palm. My uncles worked in this way and removed the sago heart. I gave a bamboo of hardened sago to my uncle, and he took a stone knife and cut the sago heart. My big brother Paipe ate the heart, and my big brother Hainai ate the heart. Surupe broke firewood with a stone knife. My father removed some tree bark with a stone knife, which he wanted to put in the house. I sat down and ate the heart of the sago palm, which my uncle had cut out. I sat down on the sago frond and I ate the sago heart.

My uncle Anapi removed the outer skin of the sago palm. With Epiae, he made the bed [to wash the sago]. Epiae went to make one as well. He went to make it for my mothers. He used a stone knife to cut the trees for this. My mother made a sago beater. She cut it with a stone knife and burnt it in the fire to make a hole.[26]

I was eating the sago shoot when planes came and flew up in the sky. My uncle was looking in the water and he saw the reflection of the planes. He did not look up at the sky. He looked at the water, and this is what he saw, and he saw a picture of what was in the sky in the water. He saw many planes flying in the sky. He said, 'Hey! There are spirit planes. There! There! They will kill us! They will kill us! They will kill us! They will kill us now! These things will kill us now! Look! Look! Look! Look! Look!' My uncle Joae said this. My uncle Anapi looked up at the sky. He looked up to the sky and saw planes flying, like *jalesopo* moths. My uncle Epiae looked up at the sky. He looked up to the sky and saw planes flying, like *jalesopo* moths. They were flying between the leaves of the trees as well, like *jalesopo* moths, but they were aeroplanes.

So my uncle Joae saw a sago palm, which had rotted. He said: 'They are spirit planes! They will kill us! They will kill us! They will kill us! They will kill us! They will kill us! Come on! Come on! Horipe's dog![27] Horipe's dog!'

He said this and he went inside [the rotten sago palm]. We went into a hole in the tree. I went inside, Paipe went inside, Jawae went inside, Anaramaniae went inside, Pua went inside, like this. Epiae went inside the hole in the tree where the others had gone. Iripe covered the opening of the hole in the tree and said, 'Come on! Horipe's dog! Huh! Come on! Horipe's dog! Huh! Come on! Horipe's dog! Huh! Come on! Horipe's dog! Huh! Come on!' he said. My uncle Anapi said this, and he looked up at the sky.

So, the aeroplanes were flying in the sky, and my father said, 'They will not kill you. Bring my children out!' he said. He said this and they still flew in the sky. We did not used to see them and this happened. He said this and we went to father's house. The mothers were making sago. The mothers were making sago and in the afternoon the planes were still flying. But the aeroplanes did not kill us. They did not kill my father and they did not kill my mother. He said, 'They have not killed ... [us]. They are just flying in the sky,' my father said, 'They are just flying in the sky'. At the time, I did not wear clothes. I wore the leaves of trees, and I killed marsupials, and I killed snakes. I went around and I used a stone knife to cut trees. I used a stone knife to cut trees and I saw this.[28]

Tipe's description of the first sighting of an aeroplane in the region is grounded in what people were doing at the time. He relates in detail the work being carried out on the sago palm, the sharing of food, the people present. These details are as important as the aeroplane sightings in his narrative. Indeed, this event can only be considered whole when the aeroplanes are placed in their interpreted context. One is not possible without the other.[29]

In addition to being a vivid evocation of what it was like to see the first aeroplanes, this is an account of sago production. It shows that an event, such as an aeroplane sighting, cannot be disembodied from the context in which it was experienced. It also gives an example of how something that might be regarded typically as 'indigenous knowledge' – the process of deriving subsistence from plants – is not just a matter of 'knowing plants'. It is a shared event, whose context is of key significance in local experience. Narratives of events in living memory are often founded on detailed evocations of their context: who was there, what work was being done, what food was shared, and so on.

On the subject of aeroplanes, Tipe's narrative shows us how an aircraft flying overhead can be a significant event in local history. Nira's account of a contemporary aeroplane landing confirms that this is still the case, even though it was a more regular occurrence in the 1990s than in the 1940s. Tipe shows that, even in the terror of perceiving something as unknown as an aeroplane, the creative appropriation and interpretation of the event, both as it happened and after its occurrence, are part of local experience. Tipe also reminds us that the aeroplane is symbolically the pinnacle of technological change. He speaks of the leaves he used to wear, in contrast to industrially manufactured clothes that arrived after the sightings of the first aeroplanes. Equally, he tells of his stone knife and the work he used to do with it, alluding to the steel tools that came later and transformed the way he worked.

Applications for development and conservation

It takes about three days of hiking for local people to navigate a path from Maimafu to Haia in the Pio-Tura region over the mountainous terrain that separates them. Language, landscape, subsistence and way of life are starkly different in the two places. In a small aeroplane, this journey takes barely ten minutes. The building of the grass airstrips in Haia and Maimafu occurred within a history of colonial contact in the wider region.[30] This commenced in the late 1800s with the first forays up the Purari River on the southern side. Patrols from the highlands first entered the region from the north in the 1930s and became more common later on. Today, both Haia and Maimafu are within an expanse of land that has been mapped out as a protected area and is now the site of an 'Integrated Conservation and Development' project. In the context of this project, 'local knowledge' is often sought. Yet interests tend to be confined to flora and fauna and human 'use' of the environment.[31] Local people are often depicted as 'threats' to biodiversity. Interactions with them are based upon project staff collecting data that satisfy externally derived goals and have perceived links with the enhancement of 'biodiversity'.

Without going into detailed ethnography of the workings of this project, we simply point out here that it would not exist without airstrips and aeroplanes. Equally, it would not exist without the interest and creative imaginings of local people. We suggest that a recognition of local history and a concerted effort to engage with it before setting up development projects would lead to greater project success and equity for local people.

There has been considerable interest in 'indigenous knowledge' as a potential 'resource' for development since the 1970s (Brokensha, Warren and Werner 1980). Twenty years later, both studies of so-called 'indigenous knowledge' and the collection of data for conservation and development projects still tend to focus on the 'use' of plants and animals and on people's interactions with 'environment'. While such studies often set out to emphasise local perspectives in development, a focus on 'knowledge' as use of the environment may not be so locally oriented or 'indigenous' after all. There is a risk that IK research represents local people in terms of categories and understandings that are deemed to be significant by protagonists of development, and not necessarily by local people themselves.

We suggest that IK research should be informed by broader anthropological concerns if the goal of more successful and equitable development is to be achieved. To this end, in this paper we have considered 'local history' as 'indigenous knowledge', from a theoretical and ethnographic standpoint. By local history in the parts of Papua New Guinea where we have worked, we do not just mean documented accounts of 'mythology', 'culture' or 'development', disembodied from experience. To emphasise this, we have focussed on oral narratives of recent and contemporary change through development. We have examined how certain women and men from the Pio-Tura region and Maimafu speak about their experiences with regard to the advent of aeroplane flights and landings above and on their land.

We have focused on these narratives of lived, remembered and performatively reworked experience in order to stress that local history (and, by extension, indigenous knowledge) is concerned with contemporary events, memory and relived experience. Understandings of local 'culture', or how local people interact with 'environment' cannot be disembodied from such experience. We suggest that, instead of restricting itself, IK research should deploy its profile in development to promote broader, more ethnographic concerns in representing local people.

We advocate both the valuation and application of anthropology and the ethnographic method in conservation and development. The specialisation of anthropological knowledge into sub-disciplines such as 'IK research' may risk a market-oriented approach aimed at furnishing conservation and development practitioners with the kinds of biological and economic analyses they favour. If this were to eclipse attempts to establish historically based understandings of the complex concerns of local people on their own terms, then 'indigenous knowledge' would be a Euro-American construct.

Notes

1   We would like to thank the following people in Papua New Guinea whose collaboration and assistance made our respective research projects and this paper possible: the people of Maimafu village; people of Haia and the Pio-Tura region; pilots and staff of Mission Aviation Fellowship, Seventh Day Adventist Aviation and New Tribes Mission; Dr. Colin Filer, Michael Laki and colleagues at the National Research Institute; and staff of the Research and Conservation Foundation. Paige West would like to especially acknowledge Pastor Les Anderson. Additionally, we thank Roy Ellen, Dorothy L. Hodgson, Jennifer Law, Christin Kocher Schmid and George Morren. We are grateful for the comments of James Leach and Andrew P. Vayda on an earlier draft of this paper and for responses to the paper by participants at the ASA Conference 2000. We acknowledge invaluable funding from the Future of Rainforest Peoples Programme (APFT) and the European Commission DGVIII and from The Wenner-Gren Foundation for Anthropological Research (Gr. 6219) and The National Science Foundation (SBR-9707719).

2   David Ellis works in Haia and the Pio-Tura region at the intersection of Simbu, Gulf and Eastern Highlands provinces with people often referred to as 'Pawaia'. Paige West works in Maimafu Village in the Lufa District of the Eastern Highlands, with people often referred to as 'Gimi'.

3   The project is known nationally and internationally as the 'Crater Mountain Wildlife Management Area'. See West (2000) and Ellis (2002) for ethnographic analysis of this project.

4   He writes: 'All manner of other acronyms are to be found in the literature for LK, such as RPK (rural people's knowledge), ITK (indigenous technical knowledge), TEK (traditional environmental knowledge) and IAK (indigenous agricultural knowledge). I prefer LK as the simplest acronym of widest currency' (Sillitoe 1998: 204).

5   It is not our intention to set this in opposition to the language of local knowledge in IK research and development. Our definition includes many of the components of 'local knowledge' outlined by Sillitoe and Posey. It is rather a heuristic device that we employ to suggest the need for a shift in approach in IK research, with practical implications for development and conservation.

6   Compare with other approaches which consider the links between human ecology and sociality, between landscape and people, nature and society – not necessarily conceived of as distinct, universal or uncontested spheres: MacCormack and Strathern (1980); Croll and Parkin (1992); Descola (1996); Descola and Pálsson (1996); Balée (1998).

7   In many respects, local history is the lifeblood of anthropology. The significance of history in ethnography has, of course, been debated extensively within anthropology from many local, regional and theoretical standpoints: see Evans-Pritchard (1962a, 1962b); Lévi-Strauss (1966); Lewis (1968); Dening (1980); Sahlins (1981); Wolf (1982); Gewertz (1983); Gewertz and Schieffelin (1985); Sahlins (1985); Vansina (1985); Lederman (1986); Cohn (1987); Gellner (1988); Hill (1988); Thomas (1989); Tonkin et al. (1989); Harrison (1990); Biersack (1991);Gewertz and Errington (1991); Sahlins (1991); Finnegan (1992); Sahlins (1992); Tonkin (1992); Foster (1995); Douglas (1996); Ingold (1996); Jorgensen (1996); Krech (1996); Knauft (1999); Lewis (1999); Borofsky (2000); and Sillitoe (2000).

8   Sillitoe emphasises 'natural resource management'. He presents local knowledge as a 'cultural' phenomenon inasmuch as 'culturally relative understanding' structures the relations people have with their environment. Posey also refers to 'natural resource management', listing it with a range of more 'biological' aspects of social life, such as nutrition, food preparation and health. He then moves on to education and social organisation in his list before returning to ecological concerns in the form of agriculture.

9   See Blum (1993) for one such example.

10   In discourses about Papua New Guinea in English, 'indigeneity' is usually expressed in terms of 'land ownership' (see Filer with Sekhran on customary land tenure, 1998: 30–31). In accordance with Brunton (e.g. 1998: 81), we prefer to use the term 'landholder', as it reflects more historical ways of relating to land which we observed as lived experience.

11   Bahuchet (1993: 14) writes of the unsuitability of the term in referring to the Congo Basin in equatorial Africa. He notes the ambiguity of the concept and the difficulty of applying it universally (ibid.: 12).

12   Gragson and Blount (1999) assert that 'ethnoecology' can only be understood in terms of, 'the knowledge, beliefs and values shared' (ibid.: viii) of each particular people. While recognition of the cogency of anthropological methods in IK research is clearly necessary, we note that research 'problems' in such cases are framed in predominantly 'biological' terms: Gragson and Blount refer to, 'localized activities directed at the satisfaction of various kinds of needs and desires,' (ibid.) as a key focus of research. Although data presented in these terms might be perceived by practitioners of conservation and development as being pertinent to the success of development projects, it is possible that both the investigation and the resulting data become a product of their own terms of enquiry and not those of local people. The phrase, 'management not just of the natural world but of humans and their culture,' (ibid.: xvii) is an example of western language, which, we suggest, has great potential to shape the conceptual framework of research.

13   Although conceptions of the category 'indigenous knowledge' appear to have broadened over the last twenty years (compare, for example, Brokensha, Warren and Werner 1980 with Warren, Slikkerveer and Brokensha 1995), there still seems to be a common correlation made between indigenous knowledge and the environment. Notable exceptions include Warren (1998) and Honey and Okafor (1998), who focus on the importance of indigenous associations, or groups, in development and indigenous knowledge, and McGovern (1999), who writes about the political and cultural implications of education systems for indigenous knowledge.

14   Consider also the treatment of indigenous knowledge, albeit slight, within international legislation (Posey 1990, 1998), including the work of the European Commission (see Braem 1999).

15   It is not our intention that 'local history' – a fluid, negotiated and evolving process – should become a commodity in development. Part of our analysis points towards a challenging of the foundations of how so called knowledges are incorporated,

appropriated and represented in development. We advocate a broadening of mechanisms in development to allow for diversity, complexity and even disagreement. We believe that anthropological research methods, if applied with a degree of independence from the constraints of development funding and time-spans, offer the potential for achieving this.

16 For discussions of compensation in relation to Papua New Guinea, see Toft (1997).

17 Representations of local people, both specifically and in general, are a central theme to each of our research projects (see West 2000; Ellis 2000, 2002). Our data are complementary, given that two distinct ethnic groups are depicted by one team of conservation biologists working on the Crater Mountain project.

18 For example, James (1995): 'For societies who own and rely on precious resources, it behoves them to modify their patterns of use in accordance with other changes taking place in their society. Rural communities in PNG should update their traditional natural resource management practices, as historical, cultural and – in an unchanging world – 'sustainable' as those practices may once have been, in order to maintain a balance with modern economic and social developments' (ibid.: 4).

19 This is not to say that data are not, in some cases, available. G. Gillison has collected extensive data on the human ecology and social practices of Gimi speaking peoples in the Eastern Highlands of Papua New Guinea (see Gillison 1993). Hide conducted an extensive study of ethnoecology in 'south Simbu', based at Karimui in the early 1980s (Hide 1984). Both Toft (1980, 1983) and Warrillow (1978, 1983) have written about the subsistence, social practices and history of Pawaia people. Egloff and Kaiku (1978, 1983) have considered sociality of Pawaia people from the point of view of prehistory in the region. The Purari River interdisciplinary studies commissioned for the hydro-electric dam (which was never built in the 1970s) provide detailed documentation on the ecology of the Upper Purari region and, to a lesser extent, on Pawaia people (Petr 1983).

20 See Morren (1986) for a discussion of 'rat' as an important food source in Papua New Guinea.

21 See Ellis (2000) for a discussion of this.

22 This place name, Maimafu, is an externally imposed demarcation. It is a way of organising local rural space and local history to meet the needs of outsiders. Although the people who live in these hamlets use it as a term to demarcate all of the hamlets as a unit for ease of communication when talking to outsiders, they do not use it when talking among themselves. They mark individual and group identity through references to men's lineages and through references to the aforementioned ridge top settlements.

23 Although the vote to rid the village of pigs took place in 1982 the actual slaughter of the last pig did not take place until an important wedding ceremony in 1984.

24 According to Joshua, a local man who was directing the overall interview, this is a metaphor for the white colour of the plane, meaning that the clouds stick to the plane and make it white.

25 There are no roads in the region, and the sounds of machines are rare.

26 The end of the sago beater has a conical hole burnt into it so that it catches the sago and scoops out the pith when it is used to beat the inside of the palm.

27 This is an old expression of alarm and excitement. Horipe's dog was apparently good at catching game.

28 This event was most probably the offensive by American bombers stationed in Port Moresby on Japanese planes at Wewak on the north coast of New Guinea on August 17th 1943 (Wagner 1979: 154). It would have involved a flight path across the centre of the country. See Wagner (ibid.: 144) for accounts of interpretations of neighbouring Daribi people about the first aeroplanes that flew overhead. For a more technical ethnography of aeroplanes, see Lemonnier (1992: 66–77).

29 This moment is also situated in the context of Tipe's childhood within his own life history. This is a short extract from a much longer account of his life.

30   See West (2000: 145–193) and Ellis (2002).
31   Local people are consulted within project 'management' procedures, but on the terms of externally designed structures, such as 'management committees'.

References

Bahuchet, S. 1993. (translated 1995). *State of indigenous populations living in rainforest areas*. Brussels: European Commission DG XI Environment.

Berlin, B., Breedlove D. E. & P.H. Raven. 1973. General principals of classification and nomenclature in folk biology. *American Anthropologist* 75: 214-242.

Bickford, D. n.d. *The villages of CMWMA*. http://www. fig.cox.miami.edu/~bickford/ villages.html.

Biersack, A. (ed.) 1991. *Clio in Oceania: Toward a historical anthropology*. Washington, DC: Smithsonian Institution Press.

Blum, E. 1993. *Making biodiversity conservation profitable: A case study of the Merck / INBio agreement. Environment* 35 (4): 16-20.

Borofsky, R. (ed.) 2000. *Remembrance of Pacific pasts: An invitation to remake history*. Honolulu: University of Hawai'i Press.

Braem, F. 1999. Indigenous peoples: In search of partners, with a consultative questionnaire. Avenir des Peuples des Forêts Tropicales *Working Paper No. 5*. Brussels: Commission Européenne/ DGVIII.

Brokensha, D., Warren, D.M. & O. Werner. (eds.) 1980. *Indigenous knowledge systems and development*. Washington, D.C.: University Press of America.

Brosius, J.P. 1999. Anthropological engagements with environmentalism. *Current Anthropology* 40 (3): 277-309.

Brunton, B.D. 1998. Private contractual agreements for conservation initiatives. In *The Motupore Conference: ICAD practitioners' views from the field: A report of the presentations of the Second ICAD Conference, Motupore Island (UPNG), Papua New Guinea, 1-5 September 1997*. (eds.) S.M. Saulei, & J-A. Ellis. Papua New Guinea: Department of Environment and Conservation/ United Nations Development Programme/ Biodiversity Conservation and Resource Management.

Bulmer, R. 1967. Why is the cassowary not a bird? A problem of zoological taxonomy among the Kalam of the New Guinea Highlands. *Man* 2: 2-25.

Chambers, R. & P. Richards. 1995. Preface. In *The cultural dimension of development: indigenous knowledge systems*. (eds.) D.M.Warren, L.J. Slikkerveer & D. Brokensha. London: Intermediate Technology Publications.

Cohn, B. 1987. *An anthropologist among the historians and other essays*. Delhi: Oxford University Press.

Conklin, H. 1954. An ethnoecological approach to shifting agriculture. *Transactions of the New York Academy of Sciences*. Series 2, 17: 133-134.

Croll, E. & D. Parkin. 1992. Anthropology, the environment and development. In *Bush base, forest farm: Culture, environment and development*. (eds.) E. Croll & D. Parkin. EIDOS. London & New York: Routledge.

Dening, G. 1980. *Islands and beaches: Discourse on a silent land, Marquesas 1774-1880*. Melbourne: Melbourne University Press.

Descola, P. 1996. *In the society of nature: A native ecology in Amazonia*. Cambridge: Cambridge University Press.

Douglas, B. 1996. Introduction: Fracturing boundaries of time and place in Melanesian anthropology. (eds.) J. Barker & D.Jorgensen. *Oceania* 66: 3, Special edition, 177-188.

Egloff, B.J. & O. Kaiku. 1978. *An archaeological and ethnographic survey of the Purari River (Wabo) dam site and reservoir, Vol. 5.* Papua New Guinea: Office of Environment and Conservation, Central Government Offices, Waigani, and Department of Minerals and Energy, P.O. Box 2352, Konedobu.

Egloff, B.J. & O. Kaiku. 1983. Prehistory and paths in the upper Purari River basin. In *The Purari - tropical environment of a high rainfall river basin.* (ed.) T. Petr. The Hague: Dr. W. Junk Publishers.

Ellis, D.M. 2000. Representations of local people by practitioners of conservation and small scale timber harvesting in Papua New Guinea. In *L'homme et la forêt tropicale, travaux de la Société d' Ecologie Humaine.* (eds.) S. Bahuchet., D. Bley., H. Pagezy & N. Vernazza-Licht. APFT. Châteauneuf de Grasse: Éditions de Bergier: 99-112.

Ellis, D.M. n.d. What is 'wildlife'? The importance of local history in development and conservation in the Pio-Tura region of Papua New Guinea. In *Custom, conservation and development in Melanesia.* (ed.) C. Filer. Canberra: Australian National University, and Port Moresby: National Research Institute of Papua New Guinea. *in press.*

Ellis, D.M. 2002. Between custom and biodiversity: local histories and market-based conservation in the Pio-Tura region of Papua New Guinea. Doctoral Dissertation, Department of Anthropology, University of Kent at Canterbury.

Ellis, J-A. 1997. *Race for the Rainforest II. Applying lessons learned from Lak to the Bismarck-Ramu Integrated Conservation and Development Initiative in Papua New Guinea.* Papua New Guinea: PNG Biodiversity Conservation and Resource Management Programme/ Department of Environment and Conservation/ United Nations Development Programme - Global Environment Facility.

Evans-Pritchard, E.E. 1962a. Social anthropology: past and present. (the Marett Lecture, 1950). In *Essays in Social Anthropology.* E.E. Evans-Pritchard. London: Faber & Faber. 13-28.

Evans-Pritchard, E.E. 1962b. Anthropology and history (1961). In *Essays in Social Anthropology.* E.E. Evans-Pritchard. London: Faber & Faber. 46-65.

Finnegan, R. 1992. *Oral traditions and the verbal arts. A Guide to research practices.* ASA Research Methods Series. London & New York: Routledge.

Filer, C. with Sekhran, N. 1998. *Loggers, donors and resource owners. Policy that works for forests and people series no.2.* Papua New Guinea. Port Moresby: National Research Institute, and London: International Institute for Environment and Development.

Foster, R.J. 1995. *Social reproduction and history in Melanesia: mortuary ritual, gift exchange, and custom in the Tanga islands.* Cambridge: Cambridge University Press.

Fowler, C. 1977. Ethnoecology. In *Ecological anthropology.* (ed.) D. Hardesty. New York: John Wiley & Sons.

Friedberg, C. 1999. Diversity, order, unity. Different levels in folk knowledge about the living. Social Anthropology (1999), 7, 1, 1-16, European Association of Social Anthropologists.

Gadgil, M. 1993. Indigenous knowledge for biodiversity conservation. *Ambio* 22 (2-3): 156.

Gellner, E. 1988. *Plough, sword and book: The structure of human history.* London: William Collins Sons & Co. Ltd.

Gewertz, D.B. 1983. *Sepik River Societies: a historical ethnography of the Chambri and their neighbours.* New Haven & London: Yale University Press.

Gewertz, D.B. & F.K. Errington. 1991. *Twisted histories, altered contexts: Representing the Chambri in a world system.* Cambridge: Cambridge University Press.

Gewertz, D.B. & E. Schieffelin. (eds.) 1985. *History and ethnohistory in Papua New Guinea.* Sydney, Australia: University of Sydney.

Gillison, G. 1993. *Between culture and fantasy: A New Guinea Highlands mythology.* Chicago and London: The University of Chicago Press.

Gragson, T.L. & B.G. Blount (eds.) 1999. *Ethnoecology: Knowledge, resources, and rights.* Athens & London: The University of Georgia Press.

Grant, N. 1996. *Community Entry for ICAD Projects – The Participatory Way.* Papua New Guinea: PNG Biodiversity Conservation and Resource Management Programme.

Harrison, S. 1990. *Stealing people's names: History and politics in a Sepik River Cosmology.* Cambridge: Cambridge University Press.

Hide, R.L. (ed.) 1984. *South Simbu: Studies in demography, nutrition and subsistence. The Research Report of the Simbu Land Use Project, Volume VI.* Boroko, Papua New Guinea: Institute of Applied Social and Economic Research, P.O. Box 5854.

Hill, J.D. 1988. *Rethinking history and myth: Indigenous South American perspectives on the past.* Urbana & Chicago: University of Illinois Press.

Hobart, M. (ed.) 1993. *An anthropological critique of development: The growth of ignorance.* London & New York: Routledge.

Honey, R. & S.I. Okafor. (eds.) 1998. *Hometown Associations: Indigenous knowledge and development in Nigeria.* London: Intermediate Technology Publications.

Ingold, T. (ed.) 1996. The past is a foreign country (1992 debate). In *Key debates in anthropology.* (ed.) T. Ingold. London: Routledge.

James, S. 1995. *The Crater Mountain Wildlife Management Area: Recommendations for developing a natural resources management plan.* New York: Wildlife Conservation Society.

Johnson, A. 1997. Processes for effecting community participation in the establishment of protected areas: A case study of the Crater Mountain Wildlife Management Area. *In The political economy of forest management in Papua New Guinea.* (ed.) C. Filer. Papua New Guinea: The National Research Institute, and London: The International Institute for Environment and Development.

Jorgensen, D. 1996. Regional history and ethnic identity in the hub of New Guinea: The emergence of the Min. Barker. In J. & Jorgensen, D. (eds.) *Oceania* 66: 189-210, Special issue.

Knauft, B.M. 1999. *From primitive to postcolonial in Melanesia and anthropology.* Ann Arbor: University of Michigan Press.

Krech, S. 1996. Ethnohistory. In *Encyclopedia of cultural anthropology, vol. 2* (eds.) D. Levinson & M. Ember. New York: Henry Holt. 558-560.

Lederman, R. 1986. Changing times in Mendi: Notes towards writing Highlands New Guinea history. *Ethnohistory* 33(1): 1-30.

Lemonnier, P. 1992. *Elements for an anthropology of technology, with a foreword by John D. Speth.* Anthropological Papers, Museum of Anthropology, University of Michigan, No. 88. Michigan: Ann Arbor.

Lévi-Strauss, C. 1966 (1962). *The savage mind.* London: Weidenfeld & Nicolson.

Lewis, I.M. (ed.) 1968. *History and social anthropology.* ASA Monograph. London: Tavistock.

Lewis, I.M. 1999. *Arguments with ethnography: Comparative approaches to history, politics and religion.* London School of Economics Monographs on Social Anthropology., Volume 70. London & New Brunswick NJ: The Athlone Press.

MacCormack, C. & M. Strathern. (eds.) 1980. *Nature, culture and gender.* Cambridge: Cambridge University Press.

McCallum, R. & N. Sekhran. 1997. *Race for the rainforest: Evaluating lessons from an integrated conservation and development 'experiment' in new Ireland, Papua New Guinea.* Papua New Guinea: PNG Biodiversity Conservation and Resource Management Programme/ Department of Environment and Conservation/ United Nations Development Programme – Global Environment Facility.

McGovern, S. 1999. *Education, modern development, and indigenous knowledge: An analysis of academic knowledge production.* New York & London: Garland Publishing.

Morren, G.E.B. 1986. *The Miyanmin: Human ecology of a Papua New Guinea Society*. Ann Arbor: UMI Research Press.

Nazarea, V.D. 1998. *Cultural memory and biodiversity*. Tucson: The University of Arizona Press.

Petr., T. (ed.) 1983. *The Purari - tropical environment of a high rainfall river basin*. The Hague: Dr. W. Junk Publishers.

Pigg, S. 1992. Inventing social categories through place: Social representation and development in Nepal. *Comparative Studies in Society and History* 34(3): 491-513.

Pigg, S. 1993. Unintended consequences: The ideological impact of development in Nepal. *South Asia Bulletin* 13 (1 & 2): 45-58.

Posey, D.A. 1990. Intellectual property rights and just compensation for indigenous knowledge. *Anthropology Today* 6 (4): August 1990.

Posey, D.A. 1998. The 'Balance Sheet' and the 'Sacred Balance': Valuing the knowledge of indigenous and traditional peoples. *Worldviews: Environment, Culture, Religion* 2 (1998): 91-106.

RCF & WCS, 1995. *Crater Mountain Wildlife Management Area: a model for testing the linkage of community-based enterprises with conservation of biodiversity (BCN Implementation Grant Proposal)*. Papua New Guinea and New York: RCF & WCS.

Richards, P. 1985. *Indigenous agricultural revolution: Ecology and food production in West Africa*. London: Unwin Hyman.

Sahlins, M. 1981. *Historical metaphors and mythical realities: Structure in the early history of the Sandwich Islands*. Ann Arbor: University of Michigan Press.

Sahlins, M. 1985. *Islands of history*. Chicago: University of Chicago Press.

Sahlins, M. 1991. *The Return of the event, again: With reflections on the beginnings of the Great Fijian War of 1843 to 1855 between the kingdoms of Bau and Rewa*. In Clio in Oceania: Towards a historical anthropology. (ed.) A. Biersack, A. (ed.) Washington: Smithsonian Institution Press. 37-99.

Sahlins, M. 1992. *Historical ethnography*. Vol. 1 of Anahulu: The anthropology of history in the Kingdom of Hawaii. (eds.) P.V. Kirch & M. Sahlins. 1992. Chicago: Chicago University Press.

Shiva, V. 1993. Women's indigenous knowledge and biodiversity conservation. In *Ecofeminism*. (eds.) M. Mies & V. Shiva. Halifax, Nova Scotia: Fernwood Publications/ London & New Jersey: Zed Books.

Sillitoe, P. 1998a. What, know natives? Local knowledge in development. *Social Anthropology* 6 (2): 203-220. European Association of Social Anthropologists.

Sillitoe, P. 1998b. The development of indigenous knowledge: A new applied anthropology. *Current Anthropology* 39 (2): April 1998.

Sillitoe, P. 2000. *Social change in Melanesia: Development and history*. Cambridge: Cambridge University Press.

Strathern, M. 1999. *Property, substance and effect: Anthropological essays on persons and things*. London & New Brunswick, NJ: The Athlone Press.

Thomas, N. 1989. *Out of time: History and evolution in anthropological discourse*. Cambridge: Cambridge University Press.

Toft, S. 1980. *A social survey of the Pawaia of the Upper Purari River: Purari River (Wabo) Hydroelectric Scheme Environmental Studies Vol. 12*. Papua New Guinea: Office of Environment and Conservation, Central Government Offices, Waigani, and Department of Minerals and Energy, P.O. Box 2352, Konedobu.

Toft, S. 1983. The Pawaia of the Upper Purari: Social aspects. *In The Purari - tropical environment of a high rainfall river basin*. (ed.) T. Petr. The Hague: Dr. W. Junk Publishers.

Toft, S. (ed.) 1997. *Compensation for resource development in Papua New Guinea. Papua New Guinea: Law Reform Commission (Monograph No. 6)*. Canberra: Resource

Management in Asia and the Pacific, Research School of Pacific and Asian Studies, The Australian National University/ Canberra: National Centre for Development Studies, The Australian National University (Pacific Policy Paper 24).

Tonkin, E. 1992. *Narrating our pasts: The social construction of oral history.* Cambridge: Cambridge University Press.

Tonkin, E., McDonald, M. & M. Chapman. (eds.) 1989. *History and ethnicity.* ASA monograph 27. London & New York: Routledge.

Vansina, J. 1985. *Oral tradition as history.* London: James Currey/ Nairobi: Heinemann.

Wagner, R. 1979. The talk of Koriki: A Daribi contact cult. *Social Research* 46: 140-165.

Warren, D. M. 1998. Foreword. In *Hometown associations: Indigenous knowledge and development in Nigeria.* (eds.) R. Honey. & S. I. Okafor. London: Intermediate Technology Publications.

Warren, D. M., Slikkerveer, L.J. & D. Brokensha. (eds.) 1995. *The cultural dimension of development: Indigenous knowledge systems.* London: Intermediate Technology Publications.

Warrillow, C. 1978. *The Pawaia of the Upper Purari (Gulf Province, Papua New Guinea): Purari River (Wabo) Hydroelectric Scheme Environmental Studies Vol. 4.* Papua New Guinea: Office of Environment and Conservation, Central Government Offices, Waigani, and Department of Minerals and Energy, P.O. Box 2352, Konedobu.

Warrillow, C. 1983. A short history of the upper Purari and the Pawaia people. In *The Purari - tropical environment of a high rainfall river basin.* (ed.) T. Petr. The Hague: Dr. W. Junk Publishers. 429-451.

Weiner, J. F. (ed.) 1988. *Mountain Papuans: historical and comparative perspectives from New Guinea fringe highlands societies.* Ann Arbor: The University of Michigan Press.

West, P. 2000. *The practices, ideologies, and consequences of conservation and development in Papua New Guinea.* Doctoral Dissertation – New Brunswick, Rutgers, the State University of New Jersey in partial fulfilment of the requirements for the degree of Doctor of Philosophy, Graduate Program in Anthropology.

Western, D., Wright, S. & S. C. Strum. (eds.) 1994. *Natural connections: Perspectives in community-based conservation.* Washington, D. C.: Island Press.

Wolf E. R. 1982. *Europe and the people without history.* Berkeley: University of California Press.

Chapter 6

The INGO, the Project, and the Investigation of 'Indigenous Knowledge': The Case of Non-Timber Forest Product (NTFP)

Sebastian Taylor

In its mainstream model of socio-economic (and ultimately urban-industrial) modernity, 'development' has, not infrequently, been indicted as the actualisation or continuation of a Western epistemological hegemony (Fowler 2000; Rist 1997; Escobar 1995; Esteva 1992). Post-modern critiques have attacked Western Enlightenment science and its assertion of an absolute, objective and singular knowledge towards the realisation of which all cultures, in the course of development, are assumed to aspire,[1] asserting instead the plurality of culture, and observing a multitude of 'sciences', of alternative epistemologies, of 'knowledges' – knowledges that are indigenous and concrete rather than universal and abstract. In the practical field of international 'development' aid, that reversal of Western emphasis has been championed by the non-government sector (here, international NGOs).

Contemporary INGO discourse privileges diversity of knowledges generated indigenously within the specific geographical and cultural localities of the people whom development is intended to serve – predominantly the subsistent, rural poor.[2] That emphasis on indigenous (or 'local') knowledge as the true measure of, and guide for, development (ascendant during and out of the 'lost decade' of 1980s and dominating structural adjustment aid policies) occurred alongside a sea-change in the kind of organisation deemed proper and best to enact aid. Where modernisation was explicitly state-led, development's 'post-modernising' period, by contrast, has demanded the agency of the non-governmental.

In this shift, development discourse has increasingly evoked (albeit in varying degrees) a kind of foundational political neutrality, situating aid beyond the state in an imperative of humanitarianism (paradoxically universal and local simultaneously) and the assertion, therein, of a consensus underlying the interaction of the Western and the indigenous. Ultimately, otherwise hostile or competitive development interests and interpretations (of Western and 'non-Western', state and non-state and so on), are implicitly brought into consensual alignment under the guiding hand of the local, the indigenous and its universal authenticity.

But there are significant problems – in this context – in the way 'indigenous knowledge' informs contemporary development theory and practice. In the first

129

place, we can ask how indigenous local knowledge and the international organisation interact. 'Indigenous knowledge', arguably, expresses itself predominantly in the immediate practices of the local group. The processes by which it becomes absorbed as knowledge through the far-flung offices of the international organisation involve, necessarily therefore, distanciation and abstraction – the conversion of actual practices into documentary record, the aggregation of incidences into general categories, and their transmission, in an essentially conceptual form, out of the local and upwards through the decision- and policy-making geo-hierarchy of the INGO. Indigenous knowledge as live practice is progressively refined (at the simplest level in the acronym, 'IK') into an abstract generalisation, in the worst instance, thereby, simply replacing the idea of the bloc of Western knowledge with an opposing but equally objectified and objectifying concept of 'the local'.

Moreover, it is arguable that 'indigenous people' rarely use a conscious category 'indigenous' in considering themselves and the practice of their lives;[3] and we can argue with greater empirical confidence that they are not always or even characteristically confined within a single comprehensively bounded social community sharing with their peers a singular, uncontested 'local' knowledge. With this in mind, the concept of 'indigeneity' appears more relevant or meaningful to those outside the sphere of the local, to be more frequently the imposition of an epistemological framework on the part of the external observer attempting to capture the meaning of a social group, its relationships and actions, in order to determine an appropriate form for its development.

IK, then, describes both the immediacy (experienced from within) of local practices and the processes by which local practice is captured and conceptualised in a documentary, generalised and abstracted category (seen from without). Indigenous knowledge is ambiguous. It is what 'they' (local people) know, and what 'we' know – or can find out – about what they know. But 'we' is itself, in the case of the international organisation, an ambiguous identity. As both 'outsider' and 'insider' (the HQ, based in the north, and the field office, based in the south), the international NGO incorporates distinct and different relations to 'the local', embodying two quite different ways of looking at 'the indigenous'. The field office, embedded in a single cultural local, sees IK primarily in specific, concrete practices that its singular locality produces. The HQ, meanwhile, overseeing multiple countries, cultural fields, field offices, and 'locals', experiences indigeneity and knowledge more distantly, as the aggregation of diverse cultural practices generalised into a global-conceptual model.

Further, where from the field office (standing in a dependent but often latently antagonistic relationship to the developing country's government), 'local' focuses on the 'community', the village, the rural, the remote – the definitively non-state, from the HQ 'local' incorporates multiple countries, in each of which both state and non-state can be described as 'local'. An INGO investigation of the local (and its IK), then, simultaneously excludes and includes state from its field of vision.

What is clear, overall, is that the powerful contemporary discourse of indigenous knowledge, in its origins in combating imposed (singular, Western) epistemology, and its influences in situating international development beyond the state, is actually characterised by ambiguity and contradiction. Through an examination of that ambiguity, it is possible to bring to light the potential of this powerful discourse to legitimise diverse and politically antagonistic developmental ends – in other words, to re-assert the underlying reality of political contest inherent in international 'development' that contemporary development discourse (including the discourse of 'the local' and indigenous knowledge) too often obscures.

Starting with the structures of the INGO and its principal development vehicle 'the project', and moving to an examination of the idea of 'forest', of 'NTFP', and the case study of an NTFP project in Laos, this chapter explores the dimension of political contest simultaneously suppressed by and embodied in the INGO discourse of indigenous 'local' knowledge.

The INGO

The international NGO is, classically, multi-sited, with offices both in the 'developed' north and the 'developing' south. Like other 'firms', the INGO has – for reasons including intelligibility to the outside world of donors, governments and 'beneficiaries', and internal imperatives of standardized practice – a clear need to conceive of itself as organisationally homogeneous and ideologically consensual. But its intrinsic multi-sitedness, most acutely in the divergence of perspectives between northern and southern offices, contains and perhaps inevitably generates very different approaches to the way in which investigation of the field of its operation, in this case 'the local', produces the organisational knowledge by which good development is produced. Stretching from the northern HQ to the southern field and project offices, the INGO comprises competing authorities and expertises of the 'outsider' and the 'insider'. Both sites – drawing on INGO discourse – emphasise 'the local' as the key determinant of 'development'. But while the southern office asserts a superior access to knowledge of the local through its proximity to specific, concrete cultural practices, the northern office asserts an alternative superiority in its overview of multiple locals and its insight (albeit paradoxical) into the common and essential characteristics of 'the local' constituted at a global level.

The interaction of north and south, across the INGO's transnational terrain, represents an organisational dynamics of understanding and knowledge in which authority (or power) over 'true development' is ambiguously distributed through an organisational space that emphasises and embodies simultaneously globality and locality. That ambiguity is intensified by a discontinuity between the INGO as organisational idea and as organisational reality. While as idea, the INGO represents fundamentally the reversal of post-war, top-down (Westernising) state-led aid, centralising, privileging and 'empowering', instead, the southern 'local', in other senses – especially with regard to money, and contracts of legal responsibility – the INGO remains representative of a more conventional hierarchy, dominated by the northern HQ.

The project

'The project' has been, and arguably remains the principal vehicle by which post-war development is converted from idea and policy into practice (Fowler 2000; Edwards & Hulme 1996; Rondinelli 1993; Madeley 1991; Oakley 1991; Lecomte 1986; Cracknell 1984; Lembke 1984; Roemer & Stern 1975).[4] Critiques of 'the project' – constituting it, like 'development', 'the North', 'the South', 'the INGO', 'the local' as if a uniform structure – attack what are seen as its hallmarks of dominance: a fixed pre-determined 'blueprint' (Fowler 2000; Madeley 1991),[5] produced out of – and reproducing in the South – an imperial Northern epistemology. Indeed, both critics and supporters of 'the project' in development draw on the same fundamental imagination of what it is – a place and an event, bounded in space and time, in which different cultural perspectives become one, either (according to critics) through the Western subordination of indigenous knowledge or (according to supporters, including INGOs) through the achievement of consensual and common knowledge (framed often in terms of an empowered and participating 'local' community).[6]

'The project' can, without doubt, be seen in bounded terms as an expression of a dominating epistemology (the quantification of villages, of villager 'participants', the frame of project time and money, the set of documents culminating in the symbolic object of the project file). But it can and should also be understood in terms of historical process and evolution, and in the context of the INGO's internally contested perspectives and organisational-epistemological dynamics, as a process of exploration. Any single project constitutes the interaction of different and diverse development knowledges. In food aid, or forest management, or health improvement, different knowledges – interpretations of local need, organisational policies, host government dispositions, logistical possibilities and constraints – all combine in the formation of the project.

The enactment of the project itself does not constitute the settling of those competing perspectives, but rather a concrete arena in which they can be tested against one another (see, e.g. Rondinelli 1993), leading to further competition in the interpretation of a project's success or failure, its impact and outcome. Projects do not simply occur as isolated events. Each is a bracketed crystallisation, as it were, rising out of the broader, embedded organisational stock of development knowledge, experience and tradition and, through the process of enactment and completion, feeding back into – and thus changing – that underlying stock.

The relationship between the INGO and its projects is a dynamic rather than a static one. The process of flow between organisational knowledge and project enactment can be characterised just as much as one that inspires contest as of one that builds consensus. More importantly, the idea that INGO 'projects' are imposed events, singular products of a uniform and dominant agency, clearly understates the degree to which contesting perspectives within the INGO interact with each other in their attempt to understand and act on the 'local' world of the to-be-developed. The remainder of the chapter will explore what happens, then, when the inherently and internally contested INGO applies the exploratory and processual medium of the project to a specific field of indigenous knowledge (forest produce). In particular it will examine what underlies the imagination of unanimity that the discourse of development led by the local promotes.

Development and 'the forest'

From the rainforest, through 'sustainable', 'social' and 'community forestry', to forest livelihoods and management, 'the forest' has become a key landscape in contemporary international development.[7] Viewed from the industrial north, forest represents a key repository – concealing an interior life from the globalising Western gaze – of the truly, authentically 'indigenous'. As such, it constitutes a prime target for contemporary INGO development and for the investigation of 'the local' and its knowledge. But if the forest is a bounded geographical site, it is equally a network of divergent and overlapping socio-economic, cultural and ecological domains. The forest is physical space, but it is also a social space. Its significance from within – in terms primarily of livelihood – is potentially quite different from its significance to the outside world. It is at one and the same time the legitimate property of 'local' people and the global inheritance of the world. It is the basis of local livelihood, it is a national resource, and it is a key 'commodity' in the international environmental debate.

It is a landscape whose developmental meaning is hotly contested. Imagination of the forest, among agents of development, contains multiple and contradictory discursive orientations – inimical constructions of 'preservation',[8] constructions of romantic (intrinsic) and economic (use) value, political claims and counter-claims of global, national and indigenous ownership. Underpinning these contesting conceptualisations lies a fundamental dichotomy: 'forest' as a legitimate lifestyle that development should acknowledge and support, or 'forest' as the aboriginal condition of subsistence that development exists to challenge and change. In this sense, the forest constitutes simultaneously the admirable sovereign autarky of local forest dwellers, and the undesirable isolation of the forest dweller from the wider human community. Set in their landscape, the foresters themselves occupy an ambivalent identity: the repository of authentic forest knowledge and thus the forest's legitimate owner-guardians, or the embodiment of unsustainable 'local' practices – trapped by the limitations of their indigenous knowledge in a juggernaut process of self-destruction. Depending on one's perspective and approach, therefore, the imperative of forest 'development' can demand – equally – the sovereignty of the forest community, and the need for the forest community to be rescued, at the limit, from itself.

'Non-timber forest product'

With the exclusion of large-scale, industrial, multinational and state-sponsored timber and logging activity from its proper 'local' sphere of influence, an alternative concept of forest, community and development has emerged in popular INGO concept and practice – one that circumnavigates 'timber' identifying, as a field of implicitly smaller-scale, more locally- or indigenously-constituted resources, 'non-timber forest products'[9] (Emerson 1997; De Beer & McDermott 1996; Peters 1996;

Fox 1995). Gradually gathered (like IK) under its own acronym,[10] 'NTFP' has become a central feature in the contemporary conceptualisation of 'local' forest knowledge and practice. While the acronym itself suggests the abstract objectification of diverse practices, co-opting them into a wider epistemological system, non-timber forest product (or 'produce') remains a congregation of diverse and contradictory elements.

'NTFP' includes: *rubber* (latexes); *firewood/fuelwood*; *rattan*; *timber*; *animals* (including both domesticated and wild, the latter including both traditional indigenous practices in hunting and trapping for skins and food, and the problems these present to the wider discourse of preservation and conservation); *edible foodstuffs* (roots, bulbs, wild crops, seeds, herbs etc.); *medicinal plants* (both for local harvest and indigenous use-practice, and for export, where, in the latter case, indigenous medicinal knowledge-practices become objectified, set against the 'modern' or Western, as 'traditional medicine'); *resins*; *aromatic woods*; *fodder/ biomass* (Emerson 1997; De Beer & McDermott 1996; Peters 1996; Fox 1995).

'NTFP' embraces both objects and practices, activities and relationships; and underlying these it constitutes diverse kinds of knowledge.[11] It includes natural growth, and tended production; direct consumption and processing; it describes the subsistent and the market-oriented, the aboriginal and the socially elaborated. It incorporates using and selling, internal self-sufficiency and the forms of contract and dependency implicit in exchange relationships. NTFP represents both the idealisation of local consumption (sustainable autarky) and the destructiveness of limited knowledge-practice. Indeed, in terms of the relationship between indigenous knowledge and forest development, it describes two fundamentally different discourses of the 'local': autonomous forest people engaged in involuted social relations of subsistent consumption; and forest people converting knowledge-objects and knowledge-practices into commodities, into income, and consequently into consumption relationships that articulate them with the wider world of market, state and, ultimately, international trade systems.

Sitting at the juncture of 'society, nature and state' (Fox 1995), NTFP offers itself differently to different development perspectives. As indigenous knowledge it can be taken, on one hand, to affirm the local and traditional, to affirm the superiority of their knowledge practices. On the other, those same indigenous knowledge practices can be constituted as the existing limits of the local/traditional, and development's key role in liberating them through integration with wider structures of modernity constituted predominantly as market and/or state. Thus, in NTFP, the process of INGO investigation actively provokes opposing constructions of indigeneity and 'development', actually generating (in spite of the state-neutrality implicit in the discourse of 'local' development) political contest.[12] On one hand, indigenous knowledge validates local resistance to external (state) incursion; on the other, IK demands the subordination of the local under external authority. This has very real consequences for the INGO, its development work, and people and communities engaged in contest with the state over ownership and management of forest resources.

The detailed analysis of an evolving project, enacted during the second half of the 1990s in Lao PDR, shows how these conceptual ambiguities, multiple forms and consequent contested truths emerge in the investigation, by different and contesting elements within the INGO, of what 'development' articulated as forest livelihood and NTFP can and should be.

The NTFP project – investigating indigenous knowledge

The following case study is taken from field research conducted in 1998 and 1999 in a large international NGO. For the purposes of anonymity, the INGO is called 'Trust International' (TI). The case-study traces the process by which different elements within the INGO – principally representatives of the northern and the southern 'fields' of European HQ (Trust Austria) and Asian field office (Trust Laos) – interact in the investigation of NTFP and the progressive formulation of an NTFP project. The study focuses particularly on an extended exchange of communications between two key INGO staff members – the Trust Austria Desk Officer (Marta) and the Trust Laos field office Director (Chuck).[13]

To start with, one may ask how Trust as an organisation becomes inspired by the idea of forest development or NTFP as a developmental medium. Consonant with INGO development discourse which privileges the 'local' (southern, indigenous), Trust International's organisational statutes identify the southern field office – authoritative in its proximity to indigenous culture – as the keystone of the organisation's global concept and practice. With this in mind, it is reasonable to expect the process of investigation that leads an INGO to 'local' knowledge and (project) practice to be significantly led by the field office. In this case-study, however, almost the reverse is the case.

In the early 1990s, the Trust Laos field office was not working directly in forestry and NTFP development. Other agencies were. Between late 1993 and early 1994, a pilot project – essentially a three-month research project – was carried out by a European firm of development consultants, 'Exsult'.[14] The project investigated the condition of forested areas and the character and conduct of non-timber forest product (NTFP) activities in one province in southern Laos. Exsult's investigation of a Laos NTFP was conducted separately from Trust, or at best in an autonomous relation to the local field office. Indeed, the early stages of that investigative research project provoked considerable animosity between Exsult and Trust Laos, where Trust fieldworkers challenged the external consultant's knowledge of (and consequently authority to describe) complex indigenous social and cultural conditions. The first interactions between local NGO office and 'external' consultant draw on the two key interpretations of IK – the authentic epistemology of a self-governing community, and the limited understanding of a people in need of development assistance. Where the external consultant espouses the latter, the INGO field office at least implicitly advances the former.

The field office's initial reaction is to attempt to discredit Exsult – to detach the external agent from the knowledge it was generating, questioning its investigative approach and findings as, in essence, oriented to interests (environmental, donor, Lao governmental) other than those of the forest 'local' and as biased towards a view of NTFP (as IK) as a problem set in wider terms of conservation, rather than a legitimate local property. Already it is possible to see here how knowledge of the local as NTFP is generated not in direct relation to the local, but between two

Investigating Local Knowledge

developmental agencies competing for conceptual (and consequently operational) legitimacy – Exsult constituting NTFP as a local problem, Trust Laos rejecting that interpretation.

Resisted by Trust Laos, Exsult looked to the original project's European donor. The donor, however, rejected Exsult's findings, and the consultant was obliged to seek an alternative audience, turning finally to the an EU government. That government's bilateral development funding agency was much more receptive, accepting and forwarding Exsult's NTFP interpretation and recommendations to the Austrian offices of Trust International. Trust Austria (in the form of desk-officer Marta) was quick to accept the donor's recommendation, recrafting Exsult's research findings as a draft project proposal and sending it, converted from initial investigation into concrete design for intervention, back to the Trust field office in Laos. The process of discovering a field of indigeneity and its conversion into a developmental proposition, conducted over the course of almost a year, finally looks like this:

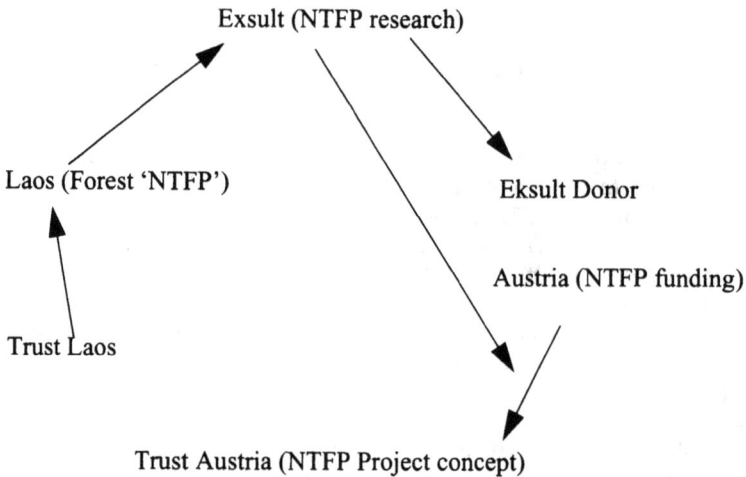

Figure 6.1 From reality to concept: the cycle of development project evolution

The most obvious missing link in this investigative chain lies between the indigenous forest people and the 'local' Trust field office.[15] An indigenous Lao forest knowledge, incorporated from the outset in terms of a pre-existing, 'NTFP' concept circulating at the global level of development discourse is, instead, 'harvested' as research by a northern consultancy, exported to Europe, equivocally evaluated by donors, recreated by Trust's European office as a project, and imported back into the local field office, characterised in that interaction as, essentially, a northern offering to a southern sibling. INGO emphasis on the dominant 'local' and its guiding knowledge notwithstanding, this 'NTFP' is not made in the south. It is, instead, made in the north. However, just as it is simplistic to suppose that the

NGO's grassroots discourse actually does create or 'empower' a truly dominant southern knowledge, so it is a mistake to assume that processes of northern epistemological-organisational cooption guarantee a dominant northern NTFP construct unchallengeable or unchallenged in the south. Rather, what follows – from the initial consultancy through to the evolution of the project – is an increasingly complex contest both among development agents for the true nature of forest people and their (NTFP-led) development, and about the true nature of indigenous knowledge and its relation to ownership, resources and political power.

The consultant's 'NTFP'

Exsult's original investigation of indigenous NTFP knowledge-practices in Laos focused on IK as the absence of an objective knowledge of the local, and saw in this the root cause of developmental problems. The research constituted foresters' knowledge-practices as, quintessentially, the developmental problem and hazard of IK which, circumscribed and limited, produces ultimately destructive local activities due to inadequate scientific and managerial regulation. The implications of this approach were that in order for regulation to be successfully imposed, those knowledge-practices needed to be captured, aggregated and submitted to external epistemological discipline. In other words, constituting indigenous Lao forest knowledge unambiguously as poorly formed or partial epistemology producing local malpractice, the research explores IK (NTFP) from the start as renegade localism in need of the centralising standardisation of the state.

Like the Lao government (for explicitly political reasons to do with controlling often separatist or state-unsympathetic forest communities), Exsult approaches the condition of indigeneity as, itself, the antithesis of 'development' – the secluded society cut off from the external world of the nation-state and beyond, hard to control, locked within limiting traditions of incomplete and progressively self-destructive knowledge. First among its research findings, the consultancy team's report concludes that shifting cultivation 'as practised by the forest communities' has been, over time, the primary cause of forest degradation. The forest – constituted in terms of the local community of knowledge and practice – is itself posed as the essential problem of, rather than the legitimate determinant of subsequent development.

The report continues: While the 'province appears to have considerable forest cover... the quality of much of the forest is less than expected'. It finds that wildlife is the 'singlemost important NTFP' but that 'as there is virtually no knowledge on the sustainability of the present hunting and trapping levels and techniques, improved management of animal populations and habitats is essential...'. The focus of the problem lands squarely on the local or indigenous population. But it is a problem constituted in a very deliberate form. The problem of indigenous forest knowledge in Laos, it suggests, is not that the local community lacks knowledge of its own practices and conditions, but that a wider (implicitly national and governmental) community lacks knowledge of them. In other words, the problem of indigenous knowledge is not in its inherent quality, but in its tendency to remain unavailable for inclusion in wider epistemological and thus regulatory structures. In Exsult's investigative approach, the 'local' is constituted as unknown. The very nature of the research, therefore, validates the existence and legitimate intervention of an outside intelligence.

Exsult's manner of looking at IK discovers a problem that demands (as 'essential') external (and disciplinary) managerial intervention. This interpretation is further refined: 'On all levels, up-to-date information and communication in relation to sustainable production of and marketing opportunities for NTFPs is lacking.' Within conditions of cooption to an external and managerial authority, NTFP is rendered even more explicitly as a matter of economic productivity (as opposed to, as argued above, the potential autarky of subsistence). In conclusion, the consultant report comes up with three recommendations: 'institutional development', 'sustainable forest management', and 'income generation', concluding that the long term objectives must be: 'To strengthen the NTFP sector[16] in Laos by reinforcing and enlarging the capacity of forest communities and the Lao government regarding the sustainable management and marketing of NTFPs; to contribute to the creation of a policy framework for the sustainable management of Laos' forest resources.'

While a reference to the 'forest communities' is included, it is positioned marginally in relation to the role an exterior, central governmental agency is expected to play in the future exploration of the indigenous and its knowledge. In the first place, NTFP is presented as a 'sector' (a level of concept conventionally assigned dominantly to state management). Moreover, where 'forest community' *is* identified, it is within the embrace of 'the Lao government'. The process of comprehending indigenous NTFP knowledge-practice is set both within the economic terms of the market and the nationalist terms of the policy framework – a policy framework explicitly geared to 'forest resources' constituted unambiguously at the level of national ownership.

Exsult's investigation of local NTFP knowledge-practices constitutes IK as primarily a detrimental condition of remote, disparate and ill-regulated localities, and the development task IK inspires as one of the aggregation of discrete knowledge-practices at an abstract level for national political and economic management. To the extent that the outsider's perspective tends to generalise specificities of IK, setting them within wider structures of social knowledge and relations, Exsult unequivocally discovers NTFP as a matter of local ignorance requiring disciplinary centralisation and state-market management.

The INGO HQ

The next stage in the exploration of NTFP and its developmental meaning is the conversion of Exsult's finding into a project concept. Trust Austria, like the consultancy firm, approaches Laos and its forest as an outsider. In parallel, it absorbs, mirrors and extends Exsult's manner of investigating and comprehending the indigenous. The European HQ reproduces Exsult's investigative lens, viewing IK (NTFP) as a matter of external, objective ignorance, deriving from the epistemological unavailability of indigenous knowledge, and the need to draw fragmented and refractory local knowledge-practices into a broader, aggregative epistemological domain. The project proposal that Trust Austria produces focuses, therefore, on the imposition on the local of (an implicitly external) standardising control.

The project's centrepiece is to be an 'NTFP Information Centre (NIC)'. Observatory-like, the 'centre' is to form the hub drawing together and distilling discrete indigenous forest knowledge-practices. Moreover, indigenous NTFP knowledge is to be constituted in terms of 'information', asserting an objectivity of knowledge over the more concrete, diverse and subjectively determined (or uncontrolled) set of knowledge-practices. The NIC is to be based in semi-official compound offices within the municipality of the capital city – quite clearly under the aegis of the government authorities. It is to comprise library and display facilities, databasing periodical publication of information, and training. These activities are themselves to be based on 'research'. Where the outsider consultant investigated NTFP as a candidate for external disciplinary management, the INGO HQ's project takes this a step further, proposing that an IK project, itself, take the form of a disciplinary investigation – the operation of a central facility in which dispersed concrete forest practices can be drawn together through the abstraction of research, converted into aggregate 'data', and reproduced through library display and documentary publication.

Like Exsult, the European HQ approaches 'the local' at a level that includes, and in fact privileges, the governmental. As argued earlier, from the international perspective, 'local' can be taken to mean indigenous characteristics seen at the national (and thus state-inclusive) level, suggesting therefore that any 'local' development automatically invokes the participation (axiomatically dominant) of the government. But more than that, Trust Austria is inclined to promote the role of the Lao government in any further investigative project, as a useful proxy through which to assert its own direct connection with Laos, in its competition with Trust Lao for access to local knowledge (where local knowledge imbues developmental authority within the INGO). As a way of emphasising this, Trust Austria's desk officer, Marta, sends the draft project proposal simultaneously to Trust Laos' field office and to the forestry department of the Lao government ensuring an approach to indigenous knowledge that offers a key role to the Lao state.

Marta's draft reads: 'The consultants say that the bilateral donor will fund the project through Trust Austria with a direct transfer of all funds to the Lao Department of Forestry (DoF).' The state-centric project is to be managerially owned by two key agencies – Trust Austria and the Lao government. Trust Laos – in INGO discourse the primary agency of 'local' interests and the true organisational repository of 'indigenous knowledge' – is totally marginalized. Indeed, in a separate communication, Marta writes to Chuck: 'I sincerely hope that you received the draft project document 'NTFP Information Center' since this would give you the possibility to meet Mr. Bounsou... who I am in contact with concerning the project.' By offering to effect an introduction between Trust Laos and a named government cadre, Marta implies the legitimacy of the state as local agent, and her own office's dominant relation to that 'local' agency.

In construing Lao locality as inclusive of the state (the level with which Western HQs generally correspond), Trust Austria gains key access to a field of local Lao knowledge (and hence authority in determining appropriate organisational development policy and response). However, this construction of local access generated through the outsider perspective sets the European HQ in direct competition with its Trust Lao colleague. In Marta's draft proposal, the field office's agency is dramatically subsidiary, allotted the unidimensional role of passive conduit channeling funds from Trust Austria to implementing governmental departments. Importantly, Trust Laos is appointed no role whatsoever in the way in which Lao NTFP is to be investigated and understood. A complex and subtle interplay of interests and perspectives between the northern and southern INGO offices ensues, involving competition over symbolic authority in project development, access to material resources projects tend to offer, but perhaps most virulently over the way in which the project itself should act as an arena for the exploration and comprehension of NTFP, indigenous knowledge and the targeted local.

The INGO 'field office'

Over the course of almost a year of project formulation between northern and southern INGO offices, the two sites become involved in contest over 'true' developmental knowledge, and knowledge of the Lao local, interwoven in a complex process of trade-offs. Marta has sent the project outline to the Lao Trust office and to the Lao government. In so doing, she has brought the international non-governmental and the state into an inextricable engagement in the pursuit of NTFP as knowledge and developmental action. Trust Austria's outsider perspective thus creates, from the outset, an ambiguous circuitry of knowledge and authority with regard to the way locality itself, and its implications for true access to local knowledge, can be defined.

Defending the field office's role as – if not substituting for – interpolating between state and 'local', immediately confronts Trust Austria's view of government as legitimate 'local' NTFP agency: 'Massive monies are handed straight over to the... government. The budget is every host government bureaucrat's dream: heavy capital expenditures, building construction, study tours to four countries, and massive training budgets.' Chuck challenges Trust Austria's 'local' authority by challenging its construction of the state-agent. He does so by invoking the classical NGO repudiation of all things governmental (Smillie, 1995) – bureaucracy, misdirection and the heavy implication of corruption. In a subsequent transmission, his bargaining for legitimate knowledge with Marta has hardened – '... we would want no role in accounting for funds transferred directly to the government'. Chuck is effectively threatening to withdraw Trust Laos' grassroots stamp of 'local' approval within the INGO as a whole, from a project where its agency is subordinated to that of the national government.

With the same hand, as it were, he offers a way out through the significant expansion of his own field office's role: 'The budget allocated to Trust Laos is insufficient to the task. The Department of Forestry is submitting the proposal and budgets this week as is to the CPC.[17] Once there, it becomes official and immutable. We have no role in implementation. The project should be redesigned with a substantial Trust Laos programming and administrative role. Otherwise the project is a joke at best.'[18] Underpinning his bid for administrative and financial project authority, Chuck asserts, this time more explicitly, his field office's superior – and pre-existent – access to and inclusion within the sphere of 'real' indigenous knowledge (the rural province and its outlying village culture, set against the urban, municipal NIC): 'Trust Laos has started a similar NTFP project in X Province.[19] We have the local knowledge and staff with language skills and management experience.'

Throughout this period, the northern HQ and southern field office are locked in competing claims of access to local knowledge. But neither's competitive epistemology focuses on 'the local' itself. 'The local', as such, is conspicuous by its absence from their communications. Instead, their inter-investigation of a Lao forest IK coalesces in a mutual exploration of the nature of the Lao state, the government, *and its relation to the local*. Chuck, on one hand, constitutes the state in classical INGO terms as top-down and bureaucratic ('official and immutable' and so on). But by utilising and reworking the ambiguous INGO identity of 'outsider' and 'insider', Marta turns this non-governmental convention on its head: 'Concerning Trust Laos' involvement... I personally believe that starting up like suggested the state-managed information centre would give the Department of Forestry and the project team a feeling of confidence, which for me is a pre-requisite for a good climate of operation. In case problems arise, things can be kept tighter later. In the past we have had very good experience with this method of 'soft' control. You might understand that any kind of top-down approach to me is not in line with nowadays development aid.'

Where Chuck had excluded the Lao state from any legitimate exploration of local knowledge (constituting the government as definitively non-local) Marta, constituting the state, from the distance of Europe, as a legitimate 'local' stakeholder, if not authority, is able to characterise that exclusion of state as an instance – on the part of the local field office – of Western, 'top-down' ascendancy. By asserting the local validity of the Lao state (implicitly locating the entire INGO, including the southern office as cultural 'other' and outsider), Marta makes the field office's claim to 'true' local knowledge authority appear a kind of mini-imperialism. Trust Laos has, itself, become the outsider (relative to the 'indigenous' Lao state). The 'local' NGO becomes the top-down outsider and the state becomes the grassroots. From whichever perspective, the investigation of the indigenous comes down to a competitive exploration, in the conduct of international development, of the nature of the state.

In a process that shows how large-scale ideological-organisational conflict is often actually played out at the level of small-scale detail and what one might call micro-discourse, Chuck and Marta pursue their contest into the narrative minutiae of the project's draft proposal. In Marta's initial offering, a sub-heading advertises 'institutional strengthening and community-based programming' as a medium of NTFP engagement and exploration (based on the intermediary role of 'local NGOs'). In a handwritten annotation next to this, Chuck writes: 'Anomaly...'. In an email to Marta, he expands: 'These objectives... are worthy but in Laos' situation anomalous based on our and other organisations' field experiences.[20] Laos already suffers from an institutional centralisation of administration that places little value on provincial development. Vientiane-based institutes/institutions have no compelling reason or record in reaching out to rural areas. In fact, such reach-out is often viewed as threatening to the central levels of self-perpetuation and existence.'

He continues: 'The weakest single programming link in the government system is "community development" or "community-based" programming. Government bodies, especially at the central, provincial and district levels, have little or no experience or understanding of these terms... Thus it is a dubious assumption for the project to assume a central level institute is to effectively implement community-based NTFP strategies and results. This is why we suggested earlier that Trust Laos have an operational, counterpart role in project implementation.'

When Marta comes back with the compromise suggestion that the Lao state could work through 'local NGOs' (maintaining Trust Laos' effective marginalisation, and emphasising the view that the international agency's field office is still not, itself, 'local'), Chuck responds by clearing the local field altogether: 'There are no indigenous/national NGOs allowed in Laos.' Any suggestion of organisational influences mediating an exploratory relationship between state and local communities – other than the local representative of the international NGO – is totally dismissed. The choice that remains is stark. You can look at indigenous knowledge-practice either through the eyes of an authoritarian state incapable of understanding the truly 'local', or through the eyes of the southern office of the INGO. Chuck goes further: 'Customary rights of local communities are being contravened by the Lao government's international logging concessions.'[21] In the field office's counter-construction, the state is not only divorced from the local, but actually stands in opposition to its interests.

The outsider's perspective sees IK as the disparate, fragmentary knowledge systems of communities, which – *left to themselves* – are environmentally and socially damaging. This perspective adopts the disciplinary tendencies of the international environmental lobby. More significantly, it supports the Lao government's contemporary nation-building project.[22] The outsiders constitute NTFP/IK as a matter for an investigative process that converts disparate knowledge-practices into centralised 'information', aggregating local NTFPs as national resource, and placing them under national ownership. Trust Laos attacks this centralist, disciplinary approach to IK, re-construing it as a form of local property that can only be understood in its physical, geo-cultural context – where IK can only truly be understood beyond the abstract generalisations of the disembodied epistemology of state information.

Chuck begins to rework the idea of the 'information centre', to challenge the dominance of the 'central', and to add in an element of decentralisation, asserting the need to understand NTFP not just as dislocated 'information' but equally as embodied, local practice. While he accedes to the material construction of the 'library'/office, he is emphatic that such a central observatory approach to the indigenous will be inadequate. He writes: 'The project will also have a significant provincial outreach role... for the first two years... followed by further field activities and research at the District level. Thus the project will have a field implementation and operations component.' Where the information-centre view construes NTFP as renegade, small-scale knowledge whose (forcible) congregation will form a wider, better forest epistemology, Chuck argues that central understanding must be rooted in an understanding of the range of decentred knowledge-practices – requiring, therefore, an intermediary agency, a requirement that Trust Laos constructs itself to fit.

More specifically, where the outsiders' perspective (like that of the state) investigates NTFP as a potential contributor to the national economy, Chuck again critiques the 'view from the centre': 'The concept of 'marketing' or selling NTFP is a real problem... There is concern that the DoF and government are taking a market, selling, and 'plantation' viewpoint to NTFPs', where the state thinks 'that the forests can be farmed and exploited for NTFPs in the same (destructive) ways in which logging has been done... The overriding fear is that the customary rights of local communities will be sacrificed'. The economic interpretation of NTFP is associated with the perspective of the state (reflected in Exsult's and then Trust Austria's approaches), and with the modernist farming project exploitatively converting indigenous practices into plantation relations.

Chuck is careful not to disavow an economic interpretation of NTFP (since one imagines, given the prevailing orthodoxy, that this would put him too far outside a viable development discourse). Instead, he writes: 'We feel that a market-oriented approach must be taken, in that no forest products should be produced for which there is not a market demand. Consumers and/or buyers and marketing channels must also be predetermined before such products should be promoted... None of us want warehouses of unsold products.' If indigenous forest knowledge-practice is to be linked up to wider systems of value and exchange, Trust Laos – as representative of the non-state – rejects the regulatory imposition of the 'predetermined' market (classically identifiable as the state marketing board), preferring instead to see local NTFP articulated with the universal economic market of supply and demand. Still we see how, in the struggle for the correct medium in which to understand and interact with IK and NTFP, the exploration remains fixed in construction and counter-construction of the validity of the role of the state.

Finally, after almost a year of north-south negotiation, the initial (Exsult/Trust Austria) draft of the project proposal (A) is replaced by a second (north-south compromise) version (B). Although represented in small editorial changes, the two documents, side by side, portray clearly the development of outsider and insider perspectives on NTFP.

Draft A sees the investigative process as one that will address the dispersed and undisciplined nature of IK, bringing discrete knowledge-practices together. Draft A thus asserts as priority the 'strong need for a focal point in Laos exclusively dealing with research, documentation, dissemination and information and training on NTFPs'. Where indigenous knowledge can itself be constituted as an objective exterior ignorance of the indigenous ('exact data on forest cover and rates of forest loss are still in short supply') *and* as local ignorance (the need for 'protection of natural resources' *from* the local population), the 'focal point' will unite those divergent knowledge-practices in conceptual categories ('research'), discipline and order them (documentation), and transmit them again, standardised, from the controlling centre outwards again as local regulation ('dissemination of information and training').

In construction of this vision of NTFP, draft A has five 'intermediate goals':

- The establishment of an NTFP Information Centre.
- Capacity-building at national level in order to strengthen the understanding of the value and potential of various NTFPs.

- Provision of a framework for action-research, policy development and application within community-based forest management plans.
- National coordination and collaboration with respect to NTFP activities in Lao PDR.
- Public awareness-raising on the environmental, social, cultural and economic value of NTFPs.

Each goal reinforces the articulation of NTFP as in essence an epistemological set to be installed, through progressive investigation and revelation, in the wider national and economic set of Lao knowledge. In this orientation, the investigation of NTFP undertaken from a central observation and aggregation point represents a kind of outward escape route for 'local communities' trapped in what is, at the centre conceived of as the self-perpetuating underdevelopment of indigeneity. Finally, in draft A, both 'applied... and adaptive research' are to be carried on at the central NIC compound in the capital city. The emphasis of the investigation – differentiated research notwithstanding – is on collectivism, collection, aggregation, and submission under a municipal standardisation of knowledge as 'information'.

By draft B, however, the central locus of 'research' has been fractured and expanded: 'Applied and adaptive research' (A) is now 'Applied and adaptive research *on-site in province as well as on Center grounds*' (B). Draft A's state/urban 'focal point' NIC, whose investigation actually draws outlying knowledge-practices into the central aggregation, has been forced to expand its physical reach, to extend its acknowledgement of NTFP outwards into the provinces. Here, we have been told, the state cannot function (and hence Trust Austria, as direct partner with state, has no tenure) without the administrative and advisorial support of Trust Laos. The exploratory terrain of the project has been expanded not only in geophysical terms, but equally in the acknowledgement of NTFP as *both* knowledge-information and knowledge-practice.

Draft A talks of 'training of staff' (retaining the central disciplinary institutional capacity-building of NTFP); draft B once again extends and expands this: 'training of staff *and selected members of participating communities*'. In draft A, the NTFP Information Centre 'will be located at the compound of the national university and the Rattan Institute'; in draft B: The NTFP Information Centre 'will be located at the compound of the national university and the Rattan Institute... however *some demonstration activities... will involve communities on the provincial level and will function as connection with theoretical research in the Center and its application for the benefit of local people*'. Draft B forces the project's epistemology to acknowledge the provincial. It weakens the dominance of the 'theoretical' approach implicit in the standardising central institution by invoking the powerful development discourse of local practice. Draft B counters draft A's upward authoritarian thrust; where A's image of the NIC elevates obscure and isolated practices into the light of aggregated, standardised, theorised knowing, B draws the project's activities of knowing back down to locales and their practices.

Conclusion

In the move away from ostensible Westernism in post-war development, the idea of other knowledges, 'indigenous' knowledges – more authentic than the Enlightenment's universal abstract in their discrete derivation from diverse conditions of society, tradition and culture around the world – has gathered considerable support.

Yet, as Bourdieu suggests, IK is an ambiguous reference. Indigenous knowledge describes local people living their daily lives to a conscious or unconscious, articulated or unarticulated epistemological framework; but it also describes the way an outsider may observe those daily lives, aggregate their individual practices, and impose as an interpretive pattern (implicitly subordinating IK under the observer's wider epistemological frame of reference). IK is both what local people know and do, and 'what local people know and do'. It asserts, simultaneously, the subjective standpoint of the local 'insider' and the objective standpoint of the 'outsider'.

The international NGO, with multiple office sites from global north to global south, engages with IK as both outsider (European HQ) and insider (Lao field office). The objective and subjective constructions of local knowledge that these organisational perspectives entail generate very different and conflicting interpretations of IK itself, and the way it should be explored and understood. The outsider sees IK as secretive, isolated, disparate and incomplete societies of knowledge cut off from the wider (national or international) epistemological community. The insider sees IK as a self-evident, authentic and inalienable process of knowledge and practice in the specific, concrete context of local society. IK is the primitive condition of 'pre-development' – the condition of limited epistemological inclusion that legitimates the intervention of the outsider's development agency; and IK is the determining context of any and all development actions – the sovereign lifeways of specific peoples.

Ultimately, the investigation of IK (here, non-timber forest product) for the purposes of creating a development project produces a project whose central purpose is to investigate IK. Most of the negotiation between the INGO's HQ and field office is actually an investigation of the nature of the project as investigative mechanism for understanding the local. And it is extremely telling that, throughout the process, the central reference point and focus of interrogation is not the local or indigenous itself, but rather the state – the Lao government – and its legitimacy (or not) as agent of local comprehension. Moreover, in attempting to explore and understand the developmental significance of the indigenous, the INGO is driven to an investigation of its own role in a knowledge made political by its multiple outsider/insider perspective.

While the INGO may attempt to broker relations between government and local in the pursuit of IK-based development, there appears to be a tendency, in the way the investigation of IK is carried out, almost inevitably to privilege the generalising, abstract and conceptual forms of indigenous knowledge, and thus to constitute the local in a way that subordinates it to centralising disciplinary processes such as those of the state. In investigating IK, the INGO does not look directly at local people or people in their localities. IK inspires, instead, contesting interpretations of the state *and its role in relation to 'the local'* leading, ultimately, to fundamental – and political – questions about the intermediary role of the non-government agency. Does INGO development work, founded in the attempt to comprehend IK, tend in any effective sense, to advance the rights of discrete groups to own and manage land according to their cultural knowledge-practices? Or does an IK approach to development tend, instead, to support, wittingly or otherwise, the very modernist developmental conventions – disciplinary epistemology, integration, and the underlying and persistent developmental infrastructure of state formation – which the post-modern or post-Western concepts of knowledge endeavoured to leave behind?

Notes

1 See, e.g. Fukuyama 1992.
2 From Rapid Rural Appraisal (RRA), through Participatory Rural Appraisal (PRA), to Participatory Learning and Appraisal (PLA) and a wide range of similar acronymous references to 'locally'-led epistemological processes, 'participation' fragments singular (Western) knowledge by privileging the knowledge of 'the other', whilst simultaneously asserting, through participation, the possibility of a singular knowledge in which Western and other knowledge become consensual.
3 Except, perhaps, when talking with visiting development professionals with whose discourse and discursive preferences they have become familiar.
4 Increasing contemporary focus on sector-level or programmatic support via state to development initiatives notwithstanding, projects continue to form a significant element in development enactment.
5 See also Ferguson 1990; Foucault 1979.
6 Though see e.g. Cooke & Kothari 2000.
7 See e.g. Hobley 1996; Fairhead & Leach 1996, 1998; ODI 1999, 2001.
8 For example, absolute maintenance of existing resources, or maintenance of resource use at replacement levels.
9 In an early incarnation referred to as 'minor forest products' (Peters 1996).
10 See, e.g. Trust International project proposal, Introductory Glossary: 'GTZ (German Development Agency)...NTFP (Non-Timber Forest Product)...MoU (Memorandum of Understanding)...UNDP (United Nations Development Programme)...'. 'NTFP' not only accrues equal status with formal institutions, organisations and interactions but indeed becomes co-opted as a form of social institution.
11 Note the fundamentally different ways in which an INGO and the World Bank interpret and define the subject matter: An INGO report describes NTFP as 'food; traditional medicine; construction and handicrafts; and wildlife' (Emerson 1997), constituting it as, primarily, a set of subjectively-determined social knowledge-practices. A World Bank report, meanwhile, describes NTFP as 'reproductive propagules; plant exudates; and vegetative structures' (Peters 1996), invoking a quite different kind of 'knowledge' based not on local practices, but on an 'objectively'-determined universal knowledge-science.
12 Contrasting with the political neutrality that INGO discourse both statutorily and often purposively avows.
13 The names have been changed.
14 Like other names, this one has been changed.
15 This part of the story was recounted to me with a mixture of amusement and chagrin by one of the senior Trust Laos expatriate staff.
16 NTFP has here been converted into its own 'sector'.

17 Central Planning Committee.
18 A key feature of INGO country office experience, knowledge and hence field-based (Southern) authority is familiarity with the 'local' history of project successes and – perhaps more pertinently – failures. It is interesting to note how, in challenging Trust Austria and the idea of the state-managed project, Chuck portrays the NIC proposal – a 'joke' – in exactly those terms that emphasise field-based knowledge, legitimacy of judgement and authority.
19 This is simply untrue; Trust Laos does, indeed, have an ongoing project involving agriculture and forestry in a separate province; however at no time is it referred to or known as an 'NTFP project'. Here, the country director is rather asserting his office's precedence in NTFP through a more generalised emphasis on the valuable NGO currency of direct grassroots experience.
20 As before, the assertion of proximity to and indeed embeddedness in local reality constituted as 'the field', with all its weight of knowledge-experience and knowledge-authority, challenges the distanced and outsider character of the Northern office.
21 Earlier, Chuck had referenced another project ongoing in Province X as proof of Trust Laos' pre-existent knowledge-experience in 'NTFP'. However, that project's proposal framed the intervention as support to the legal land-tenure rights of the state over what was constituted as 'local' and 'traditional' illegal land tenure and use; almost the exact reversal of this construction of legality in the relationship between central state and 'local' community.
22 Comprising a history of protean borders, invasion, annexation, and colonial occupation, a disruptive topography of mountains, plateaux and riverine valleys, and a population comprising somewhere between 47 and 63 ethnic groups, some of whom remain extremely recalcitrant to the very idea of a Lao national identity, the communist government in Laos has, since the 1975 'revolution', been energetic in the attempt to impose a collective framework of nation-statehood on the country.

References

Arce, A. & N. Long. (eds.) 2000. *Anthropology, Development and Modernities: Exploring discourses, counter-tensions and violence*: London & New York: Routledge.
Boli, J. & G. Thomas. (eds.) 1999. *Constructing World Culture: International nongovernmental organisations since 1875*. Stanford: Stanford University Press.
Bourdieu, P. 1990. *The Logic of Practice*. Cambridge: Polity Press.
Breman, J. 1998. The shattered image: Construction and deconstruction of the village in colonial Asia. In *Comparative Asian Studies/2*, Amsterdam: Foris Publications for the Centre for Asian Studies.
Cracknell, B. (ed.) 1984. *The evaluation of aid projects and programmes: Proceedings of the conference organized by the Overseas Development Administration at the Institute of Development Studies, University of Sussex, 7–8 April 1983*. London: ODA.
De Beer, J. & M. McDermott. 1996. *The Economic Value of Non-Timber Forest Products in Southeast Asia*. Amsterdam: Netherlands Committee for the International Union for the Conservation of Nature (IUCN).
Edwards, M. & D. Hulme. 1992. *Making a Difference: NGOs and development in a changing world*. London: Earthscan.
Emerson, B. 1997. *The natural resources and livelihood study, Ratanakiri Province, NE Cambodia*. The non-timber forest product (NTFP) project. Phnom Penh: JSRC Printing House.

Escobar, A. 1995. *Encountering Development: the making and unmaking of the Third World*. Princeton N.J.: Princeton University Press.

Ferguson, J. 1994. *The Anti-Politics Machine, Depoliticization, and Bureaucratic Power in Lesotho*. Cambridge: Cambridge University Press.

Foucault, M. 1972. *The Archaeology of Knowledge*. London: Routledge.

Foucault, M. 1977. *Discipline and Punish: The birth of the prison*. Harmondsworth: Penguin.

Fowler, A. 1997. *Striking a Balance: a guide to enhancing the effectiveness of NGOs in international development*. London: Earthscan.

Fowler, A. 2000a. NGDOs as a moment in history: Beyond aid to social entrepreneurship or civic innovation? In *Third World Quarterly: Journal of Emerging Areas* 21 (4): 637 ff. London: Carefax Publishing.

Fowler, A. 2000b. NGO futures: Beyond aid – NGDO values and the fourth position. In *Third World Quarterly: Journal of Emerging Areas* 21 (4): 589ff. London: Carefax Publishing.

Fox, J. (ed.) 1995. *Society and non-timber forest products in tropical Asia*. Occasional Papers Environment Series no. 19. Honolulu H.I.: East-West Center.

Gardner, K. & Lewis, D. 1996. *Anthropology, Development and the Post-Modern Challenge*. London: Pluto Press.

Grillo, R. & R. Stirrat. (eds.) 1997. *Discourses of Development: Anthropological perspectives*. Oxford: Berg.

Lecomte, B. 1986. *Project Aid: Limitations and alternatives*. Development Centre Studies. Paris: Development Centre of the Organisation for Economic Cooperation and Development.

Lembke, H. 1984. *Evaluating development assistance projects: Changing approaches and the conflict between scientific and administration requirements*. Occasional Paper No. 80. Berlin: German Development Institute.

Long, N. & A. Long. 1992. *Battlefields of Knowledge*: London: Routledge.

Madeley, J. 1991. *When Aid is No Help: How projects fail, and how they could succeed*. London: Intermediate Technology Publications.

Messerschmidt, D. 1995. *Rapid appraisal for community forestry: The RA process and rapid diagnostic tools*. Participatory Methodology Series, Sustainable Agriculture Programme. London: IIED.

Oakley, P. (ed.) 1991. *Projects with People: The practice of participation in rural development*. Geneva: ILO/WEP.

Peters, C. 1996. *The ecology and management of non-timber forest resources*. Technical Paper no. 322. Washington D.C.: World Bank.

Sachs, W (ed.) 1992. *The Dictionary of Development*. London: Zed Books.

Scott, J. 1990. *Domination and the Arts of Resistance: Hidden transcripts*. New Haven, Conn.: Yale University Press.

Scott, J. 1986. *Weapons of the Weak: Everyday forms of peasant resistance*. Delhi: Oxford University Press.

Smillie, I. 1995. *The Alms Bazaar: Altruism under fire – non-profit organisations and international development*. London: IT Publications.

Suzuki, N. 1998. *Inside NGOs: Learning to manage conflicts between headquarters and the field offices in non-governmental organisations*. London: Intermediate Technology Publications.

Chapter 7

Indigenous Views on the Terms of Participation in the Development of Biodiversity Conservation in Nepal

Ben Campbell

Development's curiosity with indigenous knowledge reflects perhaps the contemporary global consumer vogue for all things indigenous. While the mobile phone was the most popular gift in the U.K. for the last Christmas of the twentieth century, other popularly exchanged gifts were CDs of 'authentic' local musics from Cuba and South Africa, along with more hybrid compositions drawing on African (even 'Afro-Celtic') and Asian cultural sources. The authentic appeals of the sounds of local cultures compete with the global techno-pulse of the millennial moment. This chapter questions development's ability to follow the music industry and appropriate and consume indigenous knowledge in its appetite for new techno-ethno directions. It attempts to unpack some of the reifying consequences that can accompany seeing indigenous practice and discourse as a useful knowledge resource.

Although I argue that the terms 'indigenous' and 'knowledge' need critical qualification, it is not my intention to be dismissive of the potential for development to learn from local skills and distinctive cultural practice. Rather, I identify how a genuinely anthropological approach to knowledge-participation can involve a challenging engagement with indigenous notions of identity, power and agency, that problematizes the terms of development participation. The results might be uncomfortable for those who assume an easier project cycle choreographed to indigenous rhythms, as conflicts and contradictions are exposed that cannot effectively be ignored. How realistic is it, for instance, to extol oral knowledge in a context where modern education has become a widely promoted social goal, and village-based knowledge is structurally deprecated as backward (Pigg 1992)? Examples of problematic participation are discussed from research on conservation issues in north-central Nepal.

What is indigenous knowledge?

There are four main features of the indigenous knowledge approach identified here. Firstly, 'indigenous' need not convey the idea of a bounded, culturally specific set of coherently ordered ideas, practices and relations. It is important to work with an idea of culture that can attend to people's capacities for engaging with a diversity of truth-claiming dialogues, and to their learning processes that incorporate new skills, technologies and information.[1] Secondly, although 'indigenous' can carry connotations of 'native' or 'autochthonous', these are perhaps unnecessarily limiting of the range of groups to which the term can apply. Place rather than time – i.e. a locality-based knowledge – is a more useful grounding concept that avoids claims of residential anteriority (however politically salient these may be in some places). Thirdly, 'knowledge' may be both different from scholastic expectations of logical reflection, and greater in scope than the more 'common sense' reductions of utilitarian 'ethno-science' that do not account for symbolic cosmologies and specialist knowledges, such as of ritual practitioners. And fourthly, 'indigenous knowledge' should include practices of living that entail particular interpersonal relationships of dwelling in environments and in communities. Much subsistence know-how concerns social issues of effective group activity (i.e. participation) in coordinating and negotiating labour, residential dynamics, and gender relations, as much as it has to do with 'technical' processes and resources.

The conceptual genealogy that has generated today's coupling of the indigenous with development can be traced back through other sets of terms with different mutual relations. The primitive versus the scientific, and the traditional versus the modern are clear categorical oppositions with divergent trajectories, whereas the possibility of collaborative engagement between indigenous knowledge and development objectives suggests a blurring of distinction. There are, however, persistent contrasts commonly associated with 'the indigenous' which the term indigenous cannot itself fully express when coupled to development. Among these are, for instance, the contrasts between oral as opposed to literate cultures, between minority as opposed to dominant national ethnicities, and between livelihoods based on regional natural-resource provision as opposed to global resource circulation. Although in practice many anthropologists may be working with minority, oral, subsistence societies, the term 'indigenous' can be equally applied to literate, cash-oriented elites.[2] In strategies of ethnographic writing though, 'indigenous' is most often employed as a contrastive device, and the effects of contrast demand evaluation. 'Indigenous' is far from being a coherent analytical and comparative label in anthropology, referring to very different social realities and colonial histories when applied to continents like Asia in contrast to America or Australia (Beteille 1998, Bowen 2000).

In the context of post-colonial societies with several decades of green revolution involvement the term 'indigenous' has to convey something of the reality of hybridity between local and introduced technologies and understandings, rather than an uncontaminated, original authenticity. Akhil Gupta neatly expresses the contemporary ethnographic circumspection about identifying a distinctive indigenous terrain:

> One way to mobilize discourses of indigenous knowledge in analyzing the agricultural practices of the farmers of Alipur would have been to emphasize the use of humoral agronomy and substantivist theories. Yet this mode of analysis could not have accounted for the use of industrial inputs, the commingling of humoral accounts with bioscientific ones, or the manner in which development programs shaped farmers' agricultural decisions (Gupta 1998: 20).

For Gupta, it is hybridity rather than a dubiously nostalgic indigenousness that is a more empowering starting point for discussing the experience of the poor, the subaltern, and the marginal in South Asia. But is it necessary to argue that contexts of development intervention have brought hybrid worlds into being? Could not hybridity and mixture of rationalities be characteristic of communities less radically transformed in their eco-agronomic habits than Gupta's farmers in North India? In other words although from a post-colonial perspective the image of the indigenous appears as a coherent original tradition, there is a danger that change and diversity are thereby excluded from having a place, suggesting a static and homogeneous culture preceding development's intervention.

Himalayan hybridity

Despite representations of land-locked, otherworldly remoteness, the Himalaya has been a region of internal and external cultural traffic, an intra-continental zone of encounters and crossings. Nepal's historical position as the hub of trans-Himalayan communication was, though, seriously diminished in the previous two centuries: first by the nineteenth century Rana regime's policy of protective seclusion from British India; and second by the opening of the trade route from Calcutta direct to Lhasa via Sikkim after 1904, circumventing Kathmandu (Van Spengen 1999); and third, by China's occupation of Tibet in 1959.

Within Nepal, settlements distributed across wide altitudinal ranges have accentuated micro-differentiation of language, identity, and cultural practice to produce, with the further amplification of caste ideology, a baroque appearance of cultural diversity. The question of who are the indigenous people is not easy to answer. The idea of trying to pin down rigorous criteria for defining group A as indigenous while group B as not would be a pointless task, which in the South Asian context recalls the colonial obsession with classifying and ranking castes and tribes (Bayly 1999). Virtually all the population of Nepal claim to be descended from migrants.[3] Linguistic analysis and textual chronicles date the arrival of most of the population (both Indo-Aryan and Tibeto-Burman speakers) within the current borders of Nepal at around a thousand years ago.

Under the Panchayat system of one-party control through the monarchy (1959–90), ethnic difference was not allowed to be mobilized for political goals, it being seen as counter to the promotion of national integrity. In the last ten years, since multi-party democracy has been re-established, assertions of historical exploitation and indigenous priority by the Tibeto-Burman speaking groups have been disputed by dominant Nepali speakers, who themselves migrated in phases from parts of present-day India. But historico-mythical pasts are now being reconfigured with an eye to contemporary strategies of collective advancement and alliance formation, such as the *Janajati* federation of minorities. Claims to indigeneity look different if regions and districts are focused on, rather than considering the entire nation. The regional perspective brings out instead the history of political expansion and state formation by the dominant Nepali *Parbatiya* ethnic group since the eighteenth century. The discourse of indigenous rights has entered into Nepalese politics as a challenge to official history and the hierarchical incorporation of ethnic diversity under the caste-ordered Hindu state. While much of the organised 'indigenous' movement in Nepal is an urban and migrant phenomenon, in regionally disparate localities such as the one discussed here, significant local identities give form to differences in knowledges that are brought to bear on development processes.

Indigenous knowledge in development in Nepal

Nepal opened up to development from the 1950s. For about twenty years, between the late 1960s to late 1980s, development was predominantly concerned with addressing a population growth of over 2% per annum and its environmental consequences (Blaikie et al. 1980). Rapid deforestation by 'ignorant and fecund' peasants to make precariously terraced fields on unsuitable mountainsides was perceived as leading to disastrous soil erosion, producing massive downstream flooding and the silting up of the Bay of Bengal. This environmental crisis narrative was slowly challenged by studies that questioned the assumption of peasant ignorance as the primary cause of Himalayan environmental degradation (Ives and Messerli 1989). Not only were techniques of indigenous terrace construction re-evaluated as in fact sensibly angled for surface water run-off (Johnson et al. 1982), but historical research redistributed the blame for deforestation to include the state elite's construction of huge stucco palaces modelled on Versailles, and politicians' use of forests as bankable assets (Mahat et al. 1986).

While specialists in soil mechanics were confounded by the soundness of indigenous cultivation techniques and landslide management (Smadja 1992), Farming Systems Research in Nepal furthered appreciation of Himalayan villagers' risk-spreading practices of vertical agriculture and pastoralism, and their interest in incorporating new varieties into complexly evolving cropping regimes. Studies of local agronomic history demonstrated the ability of even relatively remote communities to intensify and diversify agronomically (Blamont 1986). Whether such processes could be said to belong to an indigenous agriculture was, though, questioned by the French anthropologist Philippe Sagant (1976), who argued that since the eighteenth century Nepal had developed a nationally uniform agricultural system of highland and lowland practices and technologies, with virtually no note of agronomic difference attributable to 'ethnic particularism'. Other comparative studies confirmed the view that there is 'no ethnic specific agriculture' in Nepal (Schroeder 1985: 35) while continuing to use the term 'indigenous' as a synonym for 'subsistence'. Leaving aside the issue of which people and which livelihood practices can be argued as 'indigenous', a recent trend among research institutions has been to focus on local agronomic particularity, such as through *in situ* seed varietal maintenance, and participatory plant breeding attending to culinary preferences for local, culturally valued strains of crop varieties (Partap and Sthapit 1998).

In regard to the forests, the move from nationalised control to community forestry spread through the 1980s. In the 1970s Fürer-Haimendorf had remarked on the destruction of Sherpa indigenous resource management systems first by forest nationalisation and later by the Sagarmatha National Park (1975). To what extent resource management systems were in fact 'indigenous' as opposed to 'traditional' was an issue raised by Bob Fisher (Gilmour and Fisher 1991) to discuss the state's coercive imposition of forest regulations via village headmen since the nineteenth century. Posing this kind of question prompts us to reflect on history, and ask from what social sources and dynamics of legitimation resource management systems have emerged. The problem of such an approach, though, is that it entails separating out practices, institutions, and roles as either internally generated and externally imposed, that have come to be collectively constitutive of hybrid contemporary regimes of environmental and political habitus.

For many years the spectre of overwhelming population growth sidelined consideration of the value indigenous knowledge could hold for development in Nepal. The urgency the issue assumed has far from disappeared, but since the late 1980s serious attempts have been made to contextualise population pressure on resources in terms of environmental justice and analysis of development policy (Blaikie & Brookfield 1987, Shrestha and Conway 1996). The extreme demographic stress on hill environments predicted in the 1970s has not materialised, and the growth of urban centres in Nepal as well as increasing patterns of out-migration to lowland Nepal, India, the Gulf and elsewhere have even reduced production intensity on hill forests in some areas. Macfarlane's (2001) brief reappraisal of demographic patterns in the village of his original study (1976), a formative text for much development policy of the time, is a sharp reminder of the danger of relying on simple Malthusian algebra for understanding processes of social, economic and environmental change. The population of Macfarlane's village did double in size by a generation later, but half of them moved out. Many of Nepal's hill villagers have moved down to the lowland Terai where, with varying entitlements to land, they have competed with commercial logging and biodiversity protected areas for access to forest margins in a more desperate struggle for land than has generally been apparent in the hills (Ghimire 1992). Shrestha and Conway's study places concerns with population growth in the context of the kinds of knowledge produced under development governmentality in Nepal, which fail to attend to the distinctive peasant ecology politics of Nepal's rural population, see Shrestha and Conway (1996).

There is, in summary, no great clarity about who is indigenous in Nepal (but much active dispute), and there is no easily identifiable way of life or knowledge practice that can be claimed as distinctively indigenous over others. The experience of rural development in Nepal has been characterised by a gradual process of learning to appreciate local knowledge in the face of failure of state directed and technologically driven formulas to relieve poverty and control population increase.

Knowing differently

In this section I try to develop an analysis of local knowledge that does not rely on the indigenous as a privileged retrojection of coherent authenticity back in time, but that gives the term 'indigenous' a perhaps surprising flexibility for attending to contesting positions of authority about knowledge and effect in the world. In terms of official census statistics the north-central district of Rasuwa, that extends up the Trisuli Valley to the border with Tibet, is virtually mono-ethnic with some 80% of its population registered as Tamang. This apparent cultural unity gives way to an internally diverse society when viewed locally. It is the coexistence of different clan identities that gives life to its communities. The incorporation of difference is expressed not in the containing endogamy of Hindu castes but in the affinal exchange of exogamous clans (bearing Tibetan derived names). Each person is made of father's bone and mother's flesh. Beef-eaters marry non-beef-eaters.[4] Buddhist lamas both contend with and complement the ritual specialisms of shamans (*bombo*) and territorial sacrificers (*lhaben*). Lamas from higher villages are considered better than those from one's own place. Lower villages grow more crops than higher villages that keep more animals. Women prefer to marry into higher up communities with healthy forests, where there will be less fodder-carrying labour for them. Potatoes from higher up make better planting tubers. Brewing yeast from lower down is more active. Products from higher locales are exchanged to mutual advantage with those from lower.[5]

These differences are not however always negotiated into happy resolutions of opposites. Tamang oral histories speak equally of conflict, combat over pasture disputes, and even warfare between intermarrying groups. But knowledge itself is regarded as one of the key areas in which differences can be best maintained. The original sacred knowledge of the world is said to have been given to two brothers in the form of books. The younger brother ate his book and became a shaman (*bombo*), speaking truth through memorised, embodied, improvised, and possessed inspiration from within. The older brother kept his book and became a lama with knowledge of the intrinsically powerful texts of the Buddha *dharma*, free from performative adulteration. The unresolved struggle between oral and literate knowledge is a defining feature of the Tamang propensity for difference. In practice, the two systems of truth co-exist as complementary to each other, rather than fighting to exert dominance. The differences are maintained by musical and ritual markers, and once initiated to become one type of specialist, a man will risk losing his mind if he dabbles in the other system. Certain types of ritual knowledge are considered inherently potent, and as they are frequently to do with unseen ghostly presences affecting our lives and bodies, they come with severe cultural health warnings. For instance, I was frequently warned off learning about curing chants as, without specialist initiation, the utterance of the words themselves was considered by lay villagers likely to make me blind.

Knowledge is appreciated as positioned and embodied. Women of certain villages have specialist knowledges of seed sowing, hat-making, yak-keeping, singing and so on. Men enjoy talking of the skills they have observed in other villages – for instance of styles of bamboo weaving, and dancing, or of activities they may have little familiarity with such as fishing. This need not mean they want to learn and adopt different knowledge. Knowledge of how others do things differently is as it were considered valuable in itself as a practice of reflection.

When development in Rasuwa District is considered, the differences of knowledge are again kept apart. The somewhat phantasmic arrival by helicopter of hundreds of apple trees to several villages about twenty years ago is illustrative. Villagers planted them as instructed but orchard maintenance and protection demanded a continuous settled presence contrary to the transhumant, agro-pastoral practices of shifting altitudinal residences and cultivation geared to vertically extensive subsistence. Fruit production, and horticultural specialisation (also promoted by development agencies) depend on a model of settled intensive farming. For lack of protection, it was not long before most of the apple trees had been destroyed by wild animals and domestic livestock. In contrast to settled intensive farming local livelihood security requires the movement of people and livestock up and down the mountainsides according to the availability of fodder, the characteristics of herd composition, the cultivation requirements of diverse crops at different elevations, and the benefits of coordinating economic and residential activities with those of other people with whom cooperation is pleasant and productive. Strategic skills, especially of gender sensitivity, are needed to maintain relations of sociability among a variety of economically interdependent clusters of herd encampments over the transhumant agro-pastoral year.

People recognise that more settled and labour intensive forms of agriculture may give increased yields, but interviews with villagers revealed that the extra manure required to increase soil fertility would work against animal health being maintained by moving beasts to different locations with a diversity of seasonal fodder species. Fodder plants are thereby better able to regenerate over the year. The alternative of using inorganic urea fertiliser has been tried by the slightly wealthier farmers, but it is seen as expensive as well as making the soil compact and difficult to work with the mattock-hoe.

The only form of development that has successfully built on indigenous knowledge of transhumant agro-pastoralism in the district is production of the famous 'yak' cheese for the tourism industry. The milch animals involved are hybrids of varying yak-cow parentage, combining altitudinal hardiness with lactational yield, and the various herding demands of the different animals put the Tamangs' ecological skills and management resourcefulness to the test. Cooperative herding arrangements add essential flexibility to household labour dynamics. The cheese factory is itself attuned to transhumant herding, as it has a mobile dairy unit that keeps close to the main concentrations of animals in the productive summer monsoon months, transporting curds back to the central unit at Shing Gombo. Cheese production is, though, at odds with many of the goals of the national park.

In sum the local knowledge of the Tamang speakers of Rasuwa is principally about living in places and communities of difference. They live between high and low altitudes, between upward and downward transhumance, between wet monsoon and dry winter, and between the vegetational poles of juniper and palm trees. Extensive movement, rather than settled intensification, is the indigenous model of productive dwelling. They live at the conjuncture of influences that they call in ritual language being 'of the middle ground' (*bar ki sa la*), that is between the historical centres of literate power in Kathmandu and Kyirong (the nearest Tibetan town).

Indigenous knowledge and biodiversity conservation

In 1976 the eastern side of the Trisuli Valley of Rasuwa District became part of the Langtang National Park, with immediate and long-term effects on local environmental practice. The park prohibited slash-and-burn cultivations, pasture management by burning, hunting for the control of crop-damaging wildlife, and unlicenced collection for use of any forest products. At this time there was no interest among conservation administrators for indigenous knowledge of the plants and animals that they saw as under threat from local villagers. The translation of nature conservation policy into everyday institutional practices of employment categories such as park rangers and game scouts, very few of whom were recruited from the local population, resulted in an interface with villagers based on evasion and entrapment. Legitimate domestic use of timber for house building was regulated by a system of licence purchasing. The high cost of licences for roof shingles made from fir trees (abies spectabilis) has led to increased use of corrugated tin, and the licence costs for the production of paper from daphne bark has meant this handicraft technology has been abandoned. Bamboo is an essential product that no farming family can do without, for mats, baskets, tethers, and rain-shields. Licences are annually procured for as many bamboo poles as a man can carry at one time, though these stocks are regularly supplemented over the year by further unlicensed and unseen collections. The park system is perceived as to do with regulation, licensing, and income generation. The enormous amount, as locals consider it, of 1,000 rupees (£10, or more than a manual worker's monthly income) is charged to each tourist for park entry.

The objectives of biodiversity conservation are simply not perceived in the interaction between park officials and villagers. It is predominantly a regime of control and punishment. Days of incarceration and negotiation of fines follow accusations of unlicensed timber collection or killing an animal such as a bear. At the same time park officials are very rarely encountered outside their offices or elsewhere than on main paths and the road to the headquarters at Dhunche. In all my many journeys through the forests of Rasuwa District, I only once met with park officials off the beaten track when a group of them were checking for unlicensed herders in high summer pastures around the cheese factory. Their dealings with these herders were frequently threatening and insulting till placatory offerings of milk or yoghurt were made, and much of the park officials' trip was spent playing cards by the firesides in herders' shelters. I questioned these officials about some of the vegetation we passed along the trails, and it was clear they had far less botanical knowledge than the villagers in the group. The park itself has no active conservation science programme, and keeps no records of important biodiversity phenomena such as the flowering of stands of different bamboo species. Villagers by contrast have good memory of these events for the six bamboo species present in the region. It has to be said in fairness that not all park officials are regarded with trepidation and disdain. There are some who show respect and compassion. The park warden at the end of the 1980s was even feted as 'a friend of the poor' for making clear to his staff that villagers did have the right to collect dead firewood for domestic use. His wife, who often wore a fur coat, was also much admired.

One of the arguments for taking an interest in indigenous knowledge of biodiversity, advocated increasingly since the 1980s under labels like 'participatory conservation', is that local or indigenous peoples have traditional concepts of oneness with the environment, or of 'kinship with the natural world' (Ramble and Chapagain 1990: 27) valuable for the goals of conservation (Müller-Böker 1995, Hay-Edie 2001). Indeed, Tamang notions of human selfhood are not radically separated off from those of other species. Clan identities in particular are seen as like natural kinds in that they bestow on their members intrinsic bone substance, but they are not species in the Western scientific sense, as they depend on making relationships with other kinds for the flesh of their reproduction. Relations between inter-marrying clans are compared to struggles between beasts (Campbell 2000) and even between the contrastive social habits of trees (Campbell 1998).

I would see a genuine indigenous knowledge of biodiversity as one that understands the range of ways in which natural species figure as both useful and meaningful to people. Tamang discourses of animal life invoke a common field of struggle between wilful agents that spills over into human relations. There is a 'phenomenological unity' (as Viveiros de Castro (1998) has written of Amerindian 'perspectivism'), across the animal-human divide, and stories of animal exploits play with interpretive exchange between animal and human characteristics. The intimacy of dwelling in such close dependence on an environment with a host of animal and plant species that provide frequent occasion for grief (e.g. crop and livestock damage, personal injury from bears, falling from trees while cutting fodder, the maddening inescapability of monsoon leeches) and joy (e.g. the pleasure of high-quality wild foods, the delight of floristic abundance celebrated in myths of cosmogenesis) is an ontology of bio-diverse connection incommensurable with modernist conservation's dichotomy of nature and society. Relating the politics of wildlife within a protected area to local cultural understandings of animals and their frequently bothersome misbehaviours raises awkward questions for advocates of the incorporation of indigenous knowledge into conservation projects. For the residents of the Langtang National Park, wildlife such as bears, wild boars, deer, monkeys, porcupines, jackals, and leopards are considered pesky gluttons of human crops and livestock. If the local perception of wildlife is as pests, a cosy image of cuddly animal lovers cannot be sustained, and in terms of the sorts of indigenous knowledge which conservation agencies are apt to pay attention to, it has to be questioned how much indulgence can be expected from non-anthropologically inclined administrators of protected areas towards such manifestly non-modern and non-conservationist natural symbolism.[6]

Arun Agrawal (1995) has forcefully argued that indigenous knowledge cannot be easily abstracted from the embedded contexts of use and meaning in which it applies, to then be used for instrumentalist development project purposes. Nor can it be reduced to a compilation of 'common sense' knowledge. During an interview I made in one of the Tamangs' mobile animal shelters, some indigenous knowledge of biodiversity was being put to use. A very pregnant buffalo had fallen and broken a leg, not an uncommon problem in this northern extent of hill-buffalo keeping. The owner's initial idea was to kill it for meat, but he was told (by a kinsman holding village political office and afraid of law-enforcers hearing about such incidents) that it is illegal to kill a pregnant buffalo in Nepal. The man had put a large saucepan on the fire, containing leaves, twigs and bark. I asked him what they were for. It was medicine for the sick buffalo he replied. When I asked about the specific plants contained in the saucepan he mentioned a story that once a man and his *mha* (sister's husband) went hunting but quarrelled after they had got their prey. They forgot about the meat that they had cut in pieces. The next day they remembered and found the meat had joined up together again. They realised the plants they had wrapped it in must be medicine. I discussed the scene later with another local friend wondering whether this medicine was known to him too. His comment was sceptical but open-minded; it could be nonsense or it could be true he told me. This medicine myth is fairly typical locally in its interplay of plant, animal and human action. It says something about Tamang understandings of the relationship between substance, conflict and knowledge, in a kind of indigenous material dialectics. It is an example of how Tamangs see struggle and contest leading to transformation, enabling new contexts for the mixture of substances to have effect. In this case the time elapsed due to the fight between affines allowed the combination of different plants to work their magic.

Whether such stories or even their tellers are listened to depends on the politics of environmental knowledge. This indigenous knowledge confronts an overall context that is not conducive to favourable 'conditions of listening' (Burghart 1996). In Nepal families with the money to do so are sending their children to English medium boarding schools to distance the next generation as much as possible from village based superstition, poverty and reliance on fields and forests for their livelihoods. Power is seen to come from science, commerce and office work, not from living close to nature (Pigg 1992). Oral knowledge in particular carries no prestige. When on different occasions I discussed my research with non-villagers (NGO workers, officials, teachers) as being to do with 'local knowledge' (which I translated into Nepali as *isthaniya bigyan* 'knowledge of place'), if they did not treat me with condescending incredulity they advised me to study the knowledge of the Buddhist lamas who, it was emphasised each time, at least had books to learn from. The distancing of oral from literate knowledge, of superstition from science, and of peasant from office worker, are markers of social power that work against the élite entertaining respect for local knowledge. The idea that scientific knowledge of environmental degradation justifies regulation of peasants' use of resources in regimes of nature conservation only increases the gulf between these contrasts of power and associated knowledge.

The storyline of nature as threatened by local people has powerful listening constituencies, especially in the alliance between international environmentalists and national park authorities in Third World states.[7] While indigenous knowledge of biodiversity is claimed to be an avenue for hearing the voice of local people who have interests in protected areas (Stevens 1997), it is a rather instrumentalized version of knowledge that is presented to environmentalists and policy makers, often in the form of lists of useful plants. If on the other hand indigenous knowledge of biodiversity is to reflect genuine cultural perceptions as anthropologists would want to explore in the round, then unfortunately for the Tamang their own mythological rather than scientific points of reference, and their antipathy to crop pests are unlikely to attract sympathy from conservationists. In the round, however, there are many ways that plants and animals are seen as vital to human life, health and proper sociality. The 'potato-thief' porcupine's quills and the Tibetan antelope's horn are essential items of shamanic curing technology, for example, and children are encouraged to adopt as pets fledgling birds fallen from nests to learn nurturing instincts. The point is that relations between species (as between clans, and castes) are characterised by engagement with diversity, manifested in a range of relationships from dependence to dispute and difficulty. The Tamang make this explicit through the course of life events, in contrast to the disengaged, de-socialised vision of nature held by conservationists.

The control over nature by the park authorities is for the most part a claim rather than a reality. The lack of adequate resourcing and poorly motivated staff keep the level of environmental surveillance to one of periodic rituals of enforcement, which assert the political relations of hierarchy between the park officials and villagers. But beyond the matter of staffing constraints in a difficult terrain, from the villagers' point of view there is another sense in which the park authorities' claim to control is flawed. This relates to local knowledge of ritual environmental legitimacy. Occasional visits by official government hunters are made for the purpose of keeping the wild boar numbers under control. They only manage to kill a few beasts at most, and leave the villagers disappointed. A young man explained why he thought the hunters were unsuccessful. He said the boars were protected by the territorial guardian of wildlife, *shyibda* (Lord of the Soil). It was as if the hunters as outsiders do not have the adequate ritual connections for permission to kill the boars. This perceived lack of adequate connectedness to the local sacred environment on the part of the park authorities underlines the problem of lack of understanding in the relationship with the local communities. The authority of the park is legitimated by the state and Western financial donors to conservation, and is enforced by the military. But it has till now little consensual participation.

How can I claim that the issues I have mentioned of ritually legitimised hunting success, and myths of medicines discovered through fighting in-laws can honestly further our understanding of indigenous knowledge of biodiversity? Hard-nosed environmental agenda-setters would presumably be dismissive, and say that what are needed are forms of knowledge that can advance the comparison of quantitative scientific indicators of changing biodiversity, such as changing percentages of forest canopy cover, and numbers of red pandas breeding. But that would be to relinquish the setting of the agenda to 'eco-crats' (Sachs 1993). Jane Guyer and Paul Richards (1996) have written about this problem in Africa, and asked how can the concept of biodiversity be framed to African needs and perspectives? They mention that it is rural communities who are often the direct custodians of biodiversity, despite what states and international agencies may think. What happens when the issue of custodianship is put on the agenda for the development of conservation policy? What can be learnt ethnographically from attempts at indigenous participation in power?

Buffer zones

The Langtang National Park was one of the first protected areas established in Nepal. The conception of the park was broadly that of the 'Yellowstone model' advocating minimum human interference within its borders. Yet Langtang is one of the most heavily populated of the parks in Nepal, and villagers were accustomed to, indeed depended on, exchanging and bartering forest produce for lowlanders' grain to help make up the average household's annual six month grain deficit. The impact of park regulations on this exchange has been hard on villagers' subsistence. A local elder statesman, who had defended the principle of the park since its inception, pointed out to me that the villagers had from the beginning only perceived the inconveniences of park regulations on their subsistence activities of wood and fodder collection, rather than appreciating the advantages such as the restriction of outsiders from using village forest resources. A revamping of the minimal human interference principle was initiated in Nepalese parks by the mid-1990s, through the buffer zone concept, piloted in Africa (Stevens 1997: 55), and was intended to give park residents legitimate access to specified areas for limited subsistence needs.

In November 1997 I visited a project intended to introduce the buffer zone principle in demonstration plots in two adjacent villages in Langtang National Park. The park had agreed to let an NGO organise the demarcation with stone walls of two sites of about one hectare each, for planting tree crops and some vegetables for the benefit of the village demonstration plot committees. However, rather than plant valued tree and plant species occurring locally such as bamboos, walnut, and wild fruit and fodder trees, the project planted mostly exotic species such as citrus. Though the villagers had been paid wages for constructing the walls, it was evident that weeding had been unsatisfactory since the plantings. Domestic livestock had also broken through the walls several times, and the plots looked as though they had received minimal attention. Discussing the situation with the NGO worker and villagers, it emerged that the villagers were primarily interested in securing as much money as possible from the NGO. They did not see the plots as meaningfully belonging to them because the park authorities had refused to discuss the villagers' main agenda, which was whether the land title to the plots would be granted back to them. Without assured ownership they considered looking after the plots a very low priority in their expenditure of time and effort, and thought the park would probably reclaim the areas after the short lifetime of the NGO's involvement. So long as some money was coming in through the project, a certain level of participation could be expected, but Tamang understandings of reciprocal advantage are far more complex and distinctive than the word 'participation' can conjure up (Campbell 1994). Standard Nepali expressions for local participation have come to be known as synonymous with unpaid, exploitative, 'voluntary' labour, evoking memories of the corvée labour system of taxation (nep. *rakam*) abused by national and local autocratic regimes in the past, as well as more recent projects to improve tourism by having villagers clean up paths and dig ditches for no immediate reward.

Visiting the Department of National Parks in the capital to enquire about the further development of the buffer zone concept for villages inside the Langtang National Park boundaries, I saw a map indicating where the buffer zone was to be. It merely covered the southern boundary of the park, and was therefore of relevance to communities outside and adjacent to the park, but ignored completely the residents inside. The model of a buffer boundary had simply been transposed from the parks in the plains area (Terai) of Nepal (specifically Chitwan and Bardia) where strict human exclusion had been instituted (Müller-Böker 1995). The map showed no appreciation of the complex transhumant use of mountain forests and pastures in seasonal movements between different altitudes, and the actual interactions of park residents with varied habitats and species. A further component of the buffer zone policy is to promise a share of 30%-50% of park income for distribution to villages that arrange to have committees and 'development plans'. The theory of participation with indigenous practice thus ends up presenting itself as an unconditional demand to follow prescribed designs for community organisation.

With the park unlikely ever to cede land title over forest areas used by villagers (the park warden refused to countenance such an event in 1997), or to match the concepts of buffer zone and the complementary idea of 'facility zone', to the range of sites actually used by the villagers, it seems that a conflict will continue between conservationist boundary maintenance of where nature and society should find their proper places, and the everyday and largely unseen practices of local people's procurements. As McNeely points out 'By... establishing national parks that have no management, the authority of governments tends to be spurious. While many governments have claimed power over resources, they lacked the capacity to implement their responsibilities, thereby creating among indigenous peoples a lack of confidence in the capacity of either state or local institutions to regulate access to local resources' (McNeely 1997: 178–179). Although the concept of buffer zone appears to invite indigenous participation to regulate resource use, it does so in a manner that requires adopting bureaucratic, committee-based procedures alien to Tamang practices of political dialogue, accountability, and dispute settlement. Indeed the establishment of national parks into remote areas has been interpreted as just such a mechanism for extending a more 'national' governmental culture into areas marked by ethnic difference from the centre (Seeland n.d.). Deeply ingrained, historical tactics of defensive recalcitrance toward central officialdom and symbolic hierarchies have been core to the persistence of indigenous vitality for the Tamang, that Holmberg (1996) characterises as a relation of cultural 'involution' against the Hindu state.[8] For policies of environmental management not to acknowledge this historical and cultural analysis is perhaps not surprising, but it provides a context for understanding uncooperative responses to 'participatory' initiatives.

The villagers' insistence on land title for the buffer zone demonstration plot, their continued practices of 'illegal' forest produce procurement, and cynicism towards offers and demands of 'participation', stem equally from the villagers' inability to handle bureaucratic process as a mechanism for their own collective strategic advantage, and the structural inability of the park authority to meet the villagers within the terms of indigenous dialogues of environmental power relations. During my fieldwork, these indigenous dialogues of power commonly took the form of notions of hunting rights and pasture use being ritually legitimised by offerings to local territorial deities, but it has to be said that more clearly political avenues for mediating community-state environmental relations had been rendered ineffective by two factors; the establishment of the park itself and the introduction of multi-party politics since 1990.

Prior to the existence of the park, village headmen (*mukhiya*) derived their authority not only from conferral of office by local district bodies, but fundamentally from their ability to coordinate village livestock movements, and to negotiate terms of pasture and forest product use by community outsiders. These headmen ensured that outsider livestock herders paid pasture fees in the form of young goats that were sacrificed in late spring and shared equally among all village households. They defended territorial boundaries from encroachment by cattle- and sheep-raiders of neighbouring communities, and declared the opening and closing of access for villagers themselves to summer and winter forest pastures, and to the open-field system after crop harvesting. When the whole context for these functions of environmental regulation were replaced by the park system, the pivotal role of the headman in managing key aspects of village productive economy was rendered impotent. Further destabilisation of village authority structures occurred with the introduction of competitive multi-party politics, and its consequences of a more individualistic pursuit of agro-pastoral strategies. Factional squabbles in this transitional period resulted in the occasional reporting of individuals for infringements of park regulations for directly political motives, though by 1998 I was told villagers had agreed upon a policy of collective silence regarding park infringements. There was not, though, much consensual basis for a proactive negotiation with park authorities on issues like compensation for crop damage by wild animals or the formation of a village management plan committee.

The story is similar in many ways to the situation recorded by Stevens (1993) of the effects of the creation of the Sagarmatha national park in the Everest region, where Sherpas' local resource use institutions were circumvented by park regulations. While locally accountable, though not necessarily 'sustainable', systems of control had been displaced, the park system was ineffective in applying its mandate of forest protection. Stevens argues that many of the Sherpa resource practices were not indigenous in the sense of being generated independently of state agency, but the point is they were familiar, and the park's blanket approach to protection was clearly insensitive to the Sherpas' own localised practices of strict protection in specified areas. Stevens mentions the cases of four villages where the institution of *shinggi nawa* was revived as a more effective means of local forest protection than the infrequent park patrols provided. His assessment is that future disagreements in resource management will continue as the park holds different goals from the locals. He says 'it may have been wiser to build on local management institutions to begin with rather than to undermine them for nearly twenty years and then attempt to reverse direction' (ibid: 326). He suggests coercive forest protection does not help win over support for conservation ideals in the long term.

Conclusion

Whether indigenous knowledge, as I have attempted to characterize it can be taken on board as relevant to nature conservation by institutions such as the Langtang National Park is doubtful. Internal Tamang discourses of power involve engagement with explicit social difference through Dravidian models of group alliance, with principles of mythologically derived creative conflict, and with dialogues across natural types. The trouble is that the park in-comers have little desire to enter the danger-zone of negotiating mutual identities bilaterally. Their authority derives precisely from originating outside the indigenous model of isogamous bilateral exchange. Perhaps when the identities and agents involved in human-environmental interaction are recognised as legitimately conflictual the debate over biodiversity can truly begin, and this will start from the basis of desire for mutual relationships between different qualities of nature. The Tamangs of the Langtang National Park are not familiar with the scientific discourse on nature conservation, and so are unable to engage conceptually with the issues raised. What they do have is an ecology of self that celebrates engagement with natural difference, which arguably resonates far more with Himalayan biodiversity than an imposed categorical distancing of society from nature, and they have an explicit language for problematizing the basis of participation, reciprocity and legitimate hierarchy in society.

The Tamangs' 'indigenous' symbolic and practical phenomenological unity between humans, territory, and species diversity runs counter to the primary feature of the environmentalist world view, which is that global biodiversity can only be saved by formalising boundaries between humanity and non-human nature (Descola 1996). 'Nature', as ascribed by Protected Area status, constitutes an unpromising project for participation because of the disruption it does to patterns of socio-biotic connection, exchange and reciprocity, or 'mediation' (Latour 1993) in lived worlds. Productive engagement with and modification of processes of growth and species interaction constitute a fundamental subsistence ontology of belonging and agency for montane agro-pastoralists. Conservation and development projects have failed to address the fundamental vertical transhumance framework of indigenous knowledge in the Langtang National Park, except for the case of the cheese factory. Regulations that prohibit the deployment of local knowledge in managing dispersed village/ forest/pasture boundary ecologies in the interest of protection destabilise the fragile viability of marginal livelihoods, the coherence of community-based leadership structures, *and* the hold people have on an understanding of the world that they do not see as polarised between nature and society.

Advocating participation with local communities in biodiversity conservation needs to address the extent to which local people's environmental agency is being challenged in the process. The example of the park buffer zone trial indicates that participatory approaches can throw up issues of profound power differences in even establishing what there is to participate about, and lack of clarity about the possible outcomes of participation, which cannot be easily side-stepped. Arun Agrawal's thoughtful contribution to the discussion on indigenous knowledge makes similar points: 'advocates of indigenous knowledge seldom emphasise that significant shifts in existing power relationships are crucial to development' (1995: 416). And further: 'It might be more helpful to frame the issue as one that requires modifications in political relationships that govern interactions between indigenous or marginalised populations, and elites or state formations' (1995: 431). 'Equitable negotiation' (Sillitoe 1998: 206) would indeed be the demand made by the residents of the national park, but belligerent non-cooperation is the more likely response as long as the terms of participation are not extended to include security of benefits beyond the lifetime of all-too-brief provisional projects experimenting in participation, and the conditions of participation – enforced bureaucratisation of village political process – skew the terms of dialogue away from indigenous negotiating practices.

Escobar has expressed scepticism about the appropriation of local knowledge of biodiversity. 'Modern biology is beginning to find local knowledge systems to be useful complements. In these discourses, however, knowledge is seen as something that exists in the "minds" of individual persons (shamans, sages, elders) about external "objects" (plants, species), the medical or economic "utility" of which their bearers are supposed to "transmit" to the modern experts. Local knowledge is not seen as a complex cultural construction, involving not objects but movements and events that are profoundly historical and relational' (1995: 204). I have tried to show how understanding the wholly different ways such knowledge connects to social and cultural fields, beyond what the codification of science implies, is in fact well served by ethnographic investigation of the notion of participation itself. 'Whose knowledge?' and 'whose participation?' are questions that lead beyond development methodology to a critical analysis of people's ability to understand their livelihoods, their environments, and their dialogue within relationships of power to others.

Within current nature protection debates, there is something of a backlash against incorporating indigenous knowledge and local community interests into conservation programmes. Wilshusen et al. (2002) note that 'new protectionists' advocating a return to strict enforcement and abandonment of conservation-with-development approaches criticise participatory initiatives for their practical ineffectiveness, and for reasons of idealised projections of local people living in eco-harmony. The new protectionists argue against linking local interests to conservation because of the internal divisions of communities, their poor organisation, and the absence of anything approximating to a conservation ethos. As a consequence it is asserted there can be no expectation of local people acting to further the goals of biodiversity protection. Recognition by the new protectionists of problems with participatory approaches can be seen to concur with much of the evidence presented here, yet the conclusions drawn are wholly different. I suggest the terms for genuine participation have hardly been glimpsed, let alone put in place.

At the other end of the debate is the position of certain development practitioners whose experience of participation as a new development orthodoxy has given cause for scepticism (Cooke & Kothari 2001). Their analysis focuses on the limits of reflexive critique within participatory frameworks, tendencies for political co-option of the local, and 'continued centralization in the name of decentralization' (2001: 7). They suggest the language of empowerment masks actual objectives of managerial efficiency, including transferring project costs onto beneficiaries. Mosse's contribution in the volume looks in particular at how local knowledge ends up not modifying project models but becomes articulated by them. In a Western India participatory farming systems project '[v]illages became easily incorporated into programme work as low-status project employees, foremen, wage-labourers, and above all as clients of the project and its field-level representatives, rather than as development partners making their own investment decisions' (2001: 26). Indigenous knowledge then had little effect on the project, and Mosse argues farmers instead learnt to manipulate 'participation' as a new form of 'planning knowledge' (ibid: 44). If this example seems simply to reinstate rather than challenge existing expectations of dependence on patronage, Uma Kothari's contribution suggests that participatory method 'purifies' or normalises power into forms of self-surveillance and consensus, that do not acknowledge the circulation of power in chains, or the possibility of subverting and disrupting the participatory discourse. Her argument highlights the problem of dealing with 'messy' aspects of people's lives that do not fit into compartmentalised participatory toolboxes. Thus, 'difference will register as deviance' (2001: 148), though she offers a more positive view of circumstances that perhaps approximate to the Tamangs' response to the buffer zone: 'exclusion can be empowering and even necessary in order to challenge existing structures of domination and control' (ibid: 151).

Participatory approaches to development are being attacked from many sides. This is probably healthy. It does matter that anthropological inflexions have become noticeable in changing conceptions of development policy, particularly regarding the characteristic of wanting to know how the world is perceived from non-dominant positions. Yet, how anthropologists' insights of these positions can be translated into strategies for intervention requires intense scrutiny. Much is lost in the translation, especially the indigenous celebration of differences that give meaning and pleasure in life. To extend this chapter's opening analogy between the commodification of authentic-sounding indigenous music and that of indigenous knowledge in development, it is noteworthy that the Tamangs' idea of musical celebration usually entails a simultaneous performance by multiple groups of religious specialists and dancing circles of villagers in a collective cacophony of different beats and voices. To record these groups separately would produce marketable works of culturally recognisable forms, but the quality of the live event with its community of participating sounds moving in an out of discordance would be entirely lost.

Notes

1   Incorporating novel elements into enhanced repertoires of indigenous practice in West Africa has been excellently discussed by Richards in terms of 'creolization' (1996). The argument in this paper is rather for the ability of indigenous knowledge systems to recognise and live with different ways of knowing.

2   In his study of the origins of colonial scientific forestry in India, the historian Richard Grove makes reference to the incorporation of indigenous models of forest protection into colonial policy. By this term he means nothing more than 'of Indian origin', as it is the punitive conservation regulations of certain Maharajas he refers to (Grove 1994).

3   A rare exception are the Chepang (Rai 1985: 2).

4   The children follow the dietary taboos of the father's clan.

5   Before roughly 1950 the main vertical expression of exchange value was salt for rice. The higher you went the more salt, the lower the more rice in the ratio of barter.

6   For a recent review of anthropological treatments of people-wildlife conflicts see Knight (2000). Community Conservation and its problems for implementation in African contexts are discussed by Adams and Hulme (1998) and more generally by Ghimire and Pimbert (1998).

7   The idea that local peasant ignorance and population growth were directly responsible for Himalayan ecological degradation was effectively demonstrated to be largely mythological by the end of the 1980s (Ives and Messerli 1989), but of course it is a persistent myth.

8   Contrasting with this argument I make about the Tamang, markedly different cultural strategies of more engaged participation with central religious and political practices are noticeable for instance in accounts such as Marie Lecomte-Tilouine's of the Magar in west central Nepal: 'L'hindouisme s'est donc présenté aux Magar...comme une condition nécessaire au pouvoir politique' (1993: 319).

References

Adams, W. & D. Hulme. 1998. 'Conservation and communities: Changing narratives, policies and practices in African conservation'. *Community Conservation in Africa* Working Paper no. 4. IDPM. University of Manchester.

Agrawal, A. 1995. Dismantling the divide between indigenous and scientific knowledge. *Development and Change* 26: 413-439.

Bayly, S. 1999. *Caste, society and politics in India from the eighteenth century to the modern age*. Cambridge: Cambridge University Press.

Béteille, A. 1998. 'The idea of indigenous people'. *Current Anthropology* 39(2): 187-191.

Blaikie, P., J. Cameron & D. Seddon. 1980. *Nepal in crisis: Growth and stagnation at the periphery*. Delhi: OUP

Blaikie, P. & H. Brookfield. 1987. *Land degradation and society*. London: Methuen.

Blamont, D. 1986. Facteurs de différenciation des systèmes agro-pastoraux des hauts pays du Centre Népal. In *Les Collines du Népal Central: écosystèmes, structures sociales et systèmes agraires'*. (ed.) J. Dobremez. Paris: INRA.

Bowen, J. 2000. Should we have a universal concept of 'indigenous peoples' rights'?' *Anthropology Today* 16(4): 12-16.

Burghart, R. 1996 *The Conditions of listening: Essays on religion, history and politics in South Asia*. Delhi: OUP.

Campbell, B. 1994. Forms of cooperation in a Tamang Community of Nepal. In *Anthropology of Nepal: peoples, problems and processes* (ed.) M. Allen. Kathmandu: Mandala. 3-18.

Campbell, B. 1998. Conversing with nature: ecological symbolism in central Nepal *Worldviews* 2(2): 123-137.

Campbell, B. 2000. Animals behaving badly. In *Natural enemies: People-wildlife conflicts in anthropological perspective*. (ed.) J. Knight. 124-144. London: Routledge.

Cooke, W. & U. Kothari. 2001. *Participation: The new tyranny?* London: Zed Books.

Escobar, A. 1995. *Encountering development*. Princeton: Princeton University Press.

Fürer-Haimendorf, C. von, 1975. *Himalayan traders*. Delhi: OUP.

Ghimire, K. 1992. *Forest or farm: The politics of poverty and land hunger in Nepal*. Delhi: Oxford University Press.

Ghimire, K. & M. Pimbert. 1998. *Social change and conservation*. London: Earthscan.

Grove, R. 1994. *Green imperialism*. Delhi: OUP.

Gupta, A. 1998. *Postcolonial developments*. Durham: Duke University Press.

Guyer, J. & P. Richards. 1996. The invention of biodiversity. *Africa* 66(1): 1-13.

Hay-Edie, T. 2001. Protecting the treasures of the Earth: Nominating Dolpo as a World Heritage site. *European Bulletin of Himalayan Research* 20-21: 46-76.

Holmberg, D. 1996. Introduction to South Asian Edition of *Order in Paradox: Myth, ritual, and exchange among Nepal's Tamang*. Delhi: Motilal Banarsidass Publishers.

Ives, J.D. & B. Messerli. 1989. *The Himalayan dilemma: Reconciling development and conservation*. London: Routledge.

Johnson, K., E. Olson, E & S. Manandhar. 1982. Environmental knowledge and response to natural hazards in mountainous Nepal. *Mountain Research and Development* 2(2): 175-188.

Knight, J. 2000. (ed.) *Natural enemies:People-wildlife conflicts in anthropological perspective*. London: Routledge. 124-144.

Lecomte-Tilouine, M. 1993. *Les dieux du pouvoir: Les Magar et l'hindouisme au Népal central*. Paris: CNRS Editions.

Macfarlane, A. 1976. *Resources and population*. Cambridge: University Press.

Macfarlane, A. 2001. Sliding down hill. *European Bulletin of Himalayan Research*. 20–21.

MacNeely, J. 1997. Interaction between biological diversity and cultural diversity. In *Indigenous Peoples, Environment and Development.* International Work Group for Indigenous Affairs. Document 85. Copenhagen.

Mahat, T., Griffin, D., & K. Shepherd. 1996. Human impact on forests in the Middle Hills of Nepal. *Mountain Research and Development* 6(4): 325-334.

Mosse, D. 2001. People's knowledge, participation and patronage: Operations and representations in rural development. In *Participation: The new tyranny?* (eds.) B. Cooke & U. Kothari. London: Zed Books.

Müller-Böker, U. 1995. *Die Tharu in Chitawan: Kenntnis, bewertung undnutzung der natürlichen umwelt im südlichen Nepal.* Stuttgart: Franz Steiner Verlag.

Partap, T. & B. Sthapit. 1998. *Managingagrobiodiversity.* Kathmandu: ICIMOD.

Pigg, S. 1992. 'Inventing social categories through space: Social representations and development in Nepal'. *Comparative Studies in Society and History* 34(3): 491-513.

Rai, N. 1985. *People of the stones: The Chepangs of Central Nepal.* Kathmandu: CNAS.

Ramble, C. & C. P. Chapagain. 1990. *Preliminary notes on the cultural dimension of conservation.* Report no.10. Makalu-Barun Conservation Project Working Paper Publication Series. DNPWC/WMI.

Richards, P. 1996. Agrarian creolization: The ethnobiology, history, culture and politics of West African rice. In. *Redefining Nature.* (eds.) R. Ellen & K. Fukui. Oxford: Berg. 291-318.

Sachs, W. 1993. *Global ecology.* London: Zed Books.

Sagant, P. 1976. *Le Paysan Limbu. Sa maison et ses champs.* The Hague: Mouton.

Schroeder, R. 1985. Himalayan subsistence systems: Indigenous agriculture in Nepal. *Mountain Research and Development* 5(1): 31-44.

Seeland, K. n.d. Paper presented to a panel on The Politics of Wildlife. European Association of Social Anthropologists, Frankfurt. September 1998.

Shrestha, N.R. & D. Conway. 1996. Ecopolitical battles at the Tarai frontier of Nepal: An emerging human and environmental crisis. *International Journal of Population Geography* 2: 313-331.

Sillitoe, P. 1998. What know natives? Local knowledge in development. *Social Anthropology.* 6(2): 203-220.

Smadja, J. 1992. Studies of climatic and human impacts and their relationship on a mountain slope above Salme in the Himalayan Middle Mountains, Nepal. *Mountain Research and Development* 12(1): 1-28.

Spengen, W. van. 1999. *Tibetan border worlds.* London: Kegan Paul.

Stephens, S. 1993. *Claiming the high ground: Sherpas, subsistence, and environmental change in the highest Himalaya.* University of California Press.

Stevens, S. (ed.) 1997. *Conservation through cultural survival.* Washington: Island Press.

Thompson, M., M. Warburton, & T. Hatley. 1986. *Uncertainty on a Himalayan scale.* London: Milton Ash Editions.

Viveiros de Castro, E. 1998. Cosmological deixis and Amerindian perspectivism. *JRAI.* 4: 469-488.

Wilshusen, P., S. Brechin, C. Fortwangler, & P. West. 2002. Reinventing a square wheel: Critique of a resurgent 'protection paradigm' in international biodiversity conservation. *Society and Natural Resources.* 15(1): 17-40.

Chapter 8

Negotiating Change, Maintaining Continuity: Science Education and Indigenous Knowledge in Eastern Canada

Trudy Sable

The criteria for success in any international development project require nothing less than ensuring cultural continuity in the affected communities while promoting change. Cultural exchanges and changes have been going on for millennia, but the terms of these exchanges are crucial. These determine whether changes are fully accepted and assimilated into people's subjective realities in a meaningful way. Development projects that do not attend to cultural continuity risk what Pierre Bourdieu terms 'symbolic violence... in which instruments of knowledge...which are arbitrary are nevertheless made to appear universal and objective. Such symbolic violence has a stultifying effect upon its recipients' (Levinson and Holland 1996: 6).

The notion of cultural continuity has to do with personal and cultural identities, how truths are negotiated, how knowledge is attained, and how that relates with power. At bottom, it is about why people accept or reject change, how they accept or reject it, and why people care about what they care about. What knowledge is important for them to ensure their physical, psychological, cultural, and spiritual survival, and what fundamental truths inform that choice and sustain cultural identity? Furthermore, how do these more localised systems relate to and manoeuvre within a national or global system? For this to occur, the relationship between truth, knowledge, and power needs to be thoroughly examined to determine how people within any cultural context negotiate to find and maintain their identity within the shifting realities of their world. What are the 'rules and conventions that give it meaning' (Giroux 1983: 20)? I propose that to undertake any effective development project requires that a process of dialogue leading to a newly constructed meaning be created between those implementing the 'development project' and those being 'developed'.

This chapter concerns a Mi'kmaw First Nation's school in Eastern Canada, where the goal is to create an alternative science curriculum in which a dialogue is created within the educational system between the traditional knowledge of the Mi'kmaq and science as it is currently taught in the provincial schools of Canada. The term education is used in its broadest sense and reaches well beyond the common notion of school. Education is used here to mean how, where, and why knowledge is acquired, interpreted, passed on and applied within any given culture, whether that knowledge is indigenous or imported. It is concerned with the systems of truth that verify this knowledge.

Two aspects of cultural continuity

The chapter draws in part on two concepts of how meaning is negotiated and constructed. The first is the theory of the 'cultural production of the educated person' as presented by educational anthropologists Dorothy Holland and Bradley Levinson (Levinson et al. 1996).

> ...all cultures and social formations develop models of how one becomes a fully 'knowledgeable' person, a person endowed with maximum 'cultural capital'...while we use the 'educated person' as an analytic construct, we argue that an indigenous conception of the educated person is variably present in all known cultures and societies (Levinson and Holland 1996: 21).

To assess these models, Holland and Levinson have attempted to develop 'a critical language for understanding relations between the school, the cultural traditions of its constituent groups, and a broader political economy' (1996: 22). They attempt to 'link local and comparative perspectives by exploring how concepts of the 'educated person' are produced and negotiated between state discourses and local practices' (1996: 8).

This is important because the notion of what it means to be educated essentially means to be knowledgeable, and what it means to be knowledgeable is relative to the system of social truths that circulate throughout the culture and become embedded within its institutions. Furthermore, social and political powers are greatly influenced by knowledge. The people, who will implement changes in the fields, factories and/or communities, have to be regarded by their peers as knowledgeable to some degree, however that is defined within their culture. It is within these educational systems, whatever forms they take, that social truths are defined and negotiated, and methods employed to determine and propagate those truths.

When conducting research among the Innu of central Labrador on cultural perceptions of the landscape, I was informed that every family had their own body of stories, which they alone could relate. Furthermore, some stories could only be told at certain times of the year. These stories had the power to awaken certain conscious beings, or produce certain effects. Similarly, Anthropologist James Leach, discusses the ownership and right to knowledge among the Nekgini speakers on the Rai Coast of Papua, New Guinea. He observes that '...expressing knowledge amounts to claiming inclusion in the relationships (including those to land and spirits), which generated that knowledge... knowledge is intimately bound up with the production of persons and identity' (Leach 2000).

Second, I draw on the concept of 'emergent' culture proposed by Raymond Williams, a concept relevant to understanding cultural continuity and production (Trujillo 1996: 120). The concept of an 'emergent' culture is the view of culture in which 'new meanings and values, new practices, new significances and experiences are continually being created' (Bauman 1977: 48). The terms 'reinvention' and 'reconstruction' have also been used to describe the process of cultural production (e.g., Schneider 1998: 191; Hess 1995). Culture is not a bounded, unchanging entity fixed in time and space, but fluid and dynamic within multiple contexts (Rosaldo

1998; Clammer, this volume). Within cultures, people are not passive recipients of change, but choose what, when, where and how they will comply with or resist any knowledge (Scott 1990; Trujillo 1996). We as researchers, development officers or scientists often forget that change is a complex, two-way process.

An example of what I believe Williams is defining as emergent culture is illustrated by the revival of powwows (a traditional practice) among the indigenous communities in eastern Canada over the last decade. The indigenous students I interviewed felt these powwows helped them identify with what it meant to be 'Indian,' even though many of the dances and chants are from other indigenous cultures throughout the U.S.A. and Canada. Some students even poked fun at certain practices that were carried on at the powwows. This is not to say that there are not 'traditional' practices, values and teachings being carried on at the powwows, but the past is merging into present concerns about identity, which has to do with defining and developing their own power within current social realities. This means indigenous people are thinking about and enacting what it means to be 'Indian' in reaction and resistance to hundreds of years of colonization. Concomitantly, a number of elders have told me that they don't like the sound of the chanting done by the younger generation, or that the younger people do not understand the drum. Some elders view the hymnal singing introduced by Catholic missionaries in the 1600s as more traditional. That does not mean one generation is more Mi'kmaq than another. These elders still go to powwows to socialize, see what is happening, and sell their crafts. It simply means that identities shift, along with what people identify with, and how people shape their identity is continually revised to incorporate new realities.

Cultures are always producing themselves over and over, emerging from their historical past and merging with current realities in the most fitting way. Knowledge and relative truth are continually negotiated as their viability and applicability to survival and well-being are re-evaluated. In the language of the Mi'kmaw people of Eastern Canada the term anku'kamkewey was used for treaty. It means 'to add more to' or 'to dance further to,' – a verb, not a noun, a process not an end product. When the treaty no longer served contemporary reality, it was further danced or added to. In my own research of pre-colonial Mi'kmaw culture, I have postulated that everything was continually sung and danced into reality, much like the 'songlines' of Australian aborigines (Chatwin 1987: 72–73). This was how people came to know, to tune into, the world as manifest at that moment of existence. This was how people found their own power.

The primary site for this negotiation, the place for 'dancing further to,' is within educational contexts of given cultural or multi-cultural settings, whether in a building, the bush, fields, home or work place. Whatever the site or context, '... schools provide each generation with social and symbolic sites where new relations, new representations, and new knowledges can be formed, sometimes against, sometimes tangential to, sometimes coincident with, the interests of those holding power' (Levinson and Holland 1996: 22). It is within the educational settings of a culture that 'different models of the 'educated person' are historically produced and contested' (Levinson and Holland 1996: 22). As Mi'kmaw educator, Dr. Marie Battiste states, 'Every generation of Aboriginal parents has had to reinvent 'education' for its children.... In my generation, 'Indian education' has become a particularly adaptable site for confronting the formal contradictions besetting Aboriginal consciousness within Canada' (Battiste 1995: vii). It is within the educational setting that Aboriginal people in Canada are questioning 'What does it mean to be an Aboriginal person' (1995: vii)?

At the Association of Social Anthropologists' conference from which this book comes, a number of discussions arose around the actual definition of 'indigenous knowledge', as well as the differentiation between 'indigenous' knowledge and 'local' knowledge. To me, the question should be what is it that provides continuity amidst change and adaptation to shifting contexts? How do people relate to their world, and what truths inform those relationships and give meaning to experiences and knowledge of any sort? Then the question can be posed about the longevity, the history, of these basic truths and the knowledge systems informed by them. The Wam people of Papua New Guinea, for example, have relinquished formal exchange feasts, rituals and ceremonies as central to their social formation in the wake of colonialism. However, 'The ethos of equality and its counterpart hierarchy... are still valid, and the struggle for status and influence is still going on...' (Stephenson 1998: 147). The shift has occurred in the resources men draw upon to contest and assert authority from the traditional growing of yams, the raising of pigs, or as holding ritual secrets to having leading positions within the new political system and village organizations, or having access to monetary resources (1998: 147). The traditional ways of affirming authority continue although the context and resources for doing so have changed.

The concepts of cultural production and emergent culture fit closely with the elements of cultural continuity that are emerging in my own work. For years (since 1993) in my work I have used a model of open concentric circles to visualize the boundaries between one's self-identity, the community in which that person exists and forms a cultural identity, and the world outside the community (Figure 8.1). An outward and inward flow exists between these circles to indicate the continual journey individuals make to negotiate their place in the world. These concentric circles exist within a three-dimensional sphere that represents the historical context in which the individual, the culture, and the wider society exists. Through the middle is a spiral to indicate both cultural continuity and the emergence or new cultural norms. The spiral, a symbol found in many cultures, has the sense that we circle back around, much like the seasons, yet we never are exactly where we were before. Realities shift yet there is an historical context from which the present arose and that, in turn, shapes the future. This is in keeping with William's concept of 'emergent' culture. It is all about how people find their identity.

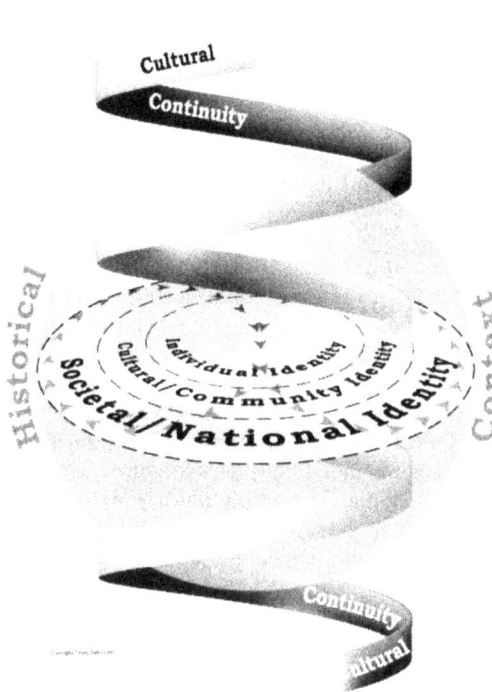

Figure 8.1   The emergent nature of culture

1. *Concentric circles* in a cross section of a sphere indicate the boundaries between one's self identity, the community in which that person exists and forms a cultural identity (which includes the school), and the world outside the community.
2. *Arrows* indicate an outward and inward flow between these circles. This flow is the continual journey any individual makes to negotiate her or his place in the world. The broken lines around the circles indicate these spheres of identity are not solid or permanent boundaries.
3. The three-*dimensional sphere* contains the concentric circles to indicate the historical context in which the individual, the culture, and the greater society exists.
4. The *spiral* through the middle indicates both *cultural continuity and the emerging nature of culture*. The spiral, a symbol found in many cultures, has the sense that we circle back around, much like the seasons, yet we never are exactly where we were before. Realities shift, as do our relationships to that reality yet there is an historical context from which the present arose and which in turn shape the future.
5.   The overall figure represents the creation of an *individual identity* as a constant generative process.

Dialogue and cultural continuity

The creation of a dialogue is the way to further cultural continuity in the context of development projects promoting change. In this context, dialogue entails a dialectical process. In anthropology, Paul Rabinow talks of the 'dialectic of fieldwork' to describe how the mutual construction of meaning occurs between 'cultural self' and 'cultural other' to create an 'intersubjective symbolic language' (Schultz & Lavenda 1995: 358). When a researcher enters an unfamiliar cultural setting, some mutual exploration and construction of meaning between that person and the people met has to occur for any communication to be effective. A process of questioning, reflecting upon one's own thinking about how things are perceived and experienced (reflexivity), and interpreting one another needs to occur until some understanding is established. Gradually, over time, 'intersections in possible ways of understanding and describing the same strip of behaviour' evolve (Schultz and Lavenda 1995: 358). Both the researcher and those within the culture, participate actively in the process, causing and affecting each other's ways of thinking.

Paolo Friere also supports dialogue in his discussion on literacy education. He believes educators must 'enter into dialogue with the people around themes that speak to the concrete situations and lived experiences that inform their daily lives...' (Giroux 1983: 227). This dialogue must be based in culturally relevant knowledge and an understanding of the ground from which people negotiate within the larger society.

A third aspect of dialogue is voiced by Henry Giroux, and is most pertinent to the attitude an anthropologist or researcher has entering any cultural setting and beginning a dialogue. In his efforts toward creating a theory of critical discourse based on dialectical thinking in education, Giroux views knowledge as embedded in a socio-political context.

> ...knowing must be seen as more than a matter of learning a given body of knowledge; it must be seen as a critical engagement designed to distinguish between essence and appearance, truth and falsity. Knowledge must not only be made problematic, stripped of its objective pretensions, it must also be defined through the social mediations and roles that provide the context for its meaning and distribution. Knowledge in this case becomes the mediator of communication and dialogue among learners (Giroux 1983:202).

In what Giroux terms 'emancipatory rationality,' he proposes we look into heart of ideology itself, using dialectical thinking to critique the limitations and rationalizations that restrict 'thinking about the process of thinking itself' (1983: 191). It is necessary 'to confront assumptions concerning the aims of education – assumptions regarding who is going to be educated, and assumptions about what kinds of knowledge, values and social relationships are going to be deemed legitimate as educational concerns' (1983: 193).

Finally, Maori spokes woman and author, Linda Tuhiwai Smith, warns that even this type of 'emancipatory rationality' has not 'freed researchers from exercising intellectual arrogance or employing evangelical and paternalistic practices' (Smith 1999: 177). She derides the underlying idealism fuelling projects conducted for the greater common good for oppressed communities (1999: 2, 24). Research '...is regulated through the formal rules of individual scholarly disciplines and scientific paradigms, and the institutions that support them (including the state). It is realized in the myriad of representations and ideological constructions of the "Other"...' (1999: 7–8). John Campbell's work (this volume) in Botswana offers an example of Smith's charge in his illustration of how the use of Geographical Information Systems can be skewed and politically implemented in mapping land use. No research exists in innocence free from political and social conditions.

Smith also recognizes the conundrum of indigenous peoples trying to protect their traditional ways while existing within dominant societies and the need 'to transform our lives on a larger scale than our own localized circumstances' (1999: 39). This struggle necessitates a 'dialogue across the boundaries of oppositions' because of the constant collision indigenous peoples encounter with dominant views in their attempts to transform their own lives (1999: 39).

The steps toward creating a dialogue are multiple, and familiar to anthropologists. To undertake research in any culture requires us to be self-reflecting about our motivations and ideals, the fundamental 'truths and falsities' we are carrying with us, and what we have as our goals. 'What are your motivations?' is a question I am continually asked when I enter indigenous Canadian communities to undertake any project. In other words, any model we use for studying 'other' must also be applied to ourselves. Even the term 'development' and ideas about what and who is being developed to what end should be scrutinized. As one Malaysian man said to me at the recent Association for Social Anthropologists conference when I asked what development meant to him, 'We simply do not want to be colonized again.' Without the ability to negotiate their own truths, find their own ground within a dialogue, people will not be receptive to proposed changes in a meaningful way.

The proposed reasons we have for change may be entirely inadequate from the developing country's point of view. This is the cause of failure for many development projects. For development projects to be sustainable means more than renewable resources; it requires a sustained dialogue and people's willingness to participate in the process of change. Even if we find mutual goals, the means for achieving those goals, the process for instituting them, and the models used are often inadequate (Edwards 1998; Agrawal & Gibson 1999). At a meeting with representatives of the Department of Indian and Northern Affairs to discuss a partnership project between Eastern Canadian indigenous communities and universities to deal with economic development, it was openly acknowledged by one representative that past attempts for economic development projects on reserves had failed. These failures stemmed from a narrow approach and inappropriate concept of what economic development entailed, neglecting the cultural underpinnings necessary for any project to be successful. They were now looking to adopt a far more comprehensive and long-term approach in partnership with indigenous communities. Without creating this sense of collectivist, the relationship becomes one of dominant and dominated, or a series of strategic manoeuvres to push development agendas. Finally we need to determine who is accountable for changes promoted.

Western science education: education or enculturation?

A growing body of literature purports that science in the west is a culture of scientists with a distinct worldview, traditions and exclusive membership (Hess 1995; Haraway 1989; Ogawa 1999; Cobern 1999; Aikenhead & Jegede, 1999; Knorr-Cetina 1981). This notion of science as culture needs to be understood as part of the self-reflection process necessary for to creating a dialogue with other cultures. This self-reflection process is vital to understanding the growing opposition among indigenous peoples regarding their rights to govern and design their own educational systems.

The Mi'kmaq consider themselves subjects of 400 years of colonisation, yet continue to assert their rights to self-government. Specifically, many indigenous peoples view western education as an imposition, a means to negate their cultural traditions. Science, with its methods of inquiry and theoretical constructs, is considered a dominant ideology that has blanketed their cultures with a claim to universality, progress, and power. Such statements as, 'The Western world view with its aggressive educational practices and technoscience orientation has placed indigenous cultures in harm's way' (Kawagley 1995: 2), or 'The greater menace to Aboriginal thinking are the assumptions that drive the search for knowledge in the Western world' (Ermine 1995: 102) echo throughout contemporary indigenous literature and rhetoric. Furthermore, ethnoscience itself has been criticized. Vine Deloria, a Standing Rock Sioux and Professor at the University of Colorado, recognizes the well-intentioned efforts of ethnoscientists. However, he warns, 'One of the difficulties today in speaking about tribal knowledge is the tendency to suggest that when traditional teachings correspond to the findings or present beliefs of western science, then traditional wisdom is validated' (Deloria 1992: 15). The clocks cannot be reversed but we can begin to listen carefully and learn well from what indigenous people throughout the world have been trying to say for centuries – it is a time for dialogue not lessons.

In general, Aboriginal children in the Maritime provinces of eastern Canada are 1–2 grade levels below their non-Aboriginal counterparts when they enter the provincial schools from reserve schools, and fall further behind by Grade 12. The majority of these students do not pass into higher levels of the science courses and graduate lacking the academic achievement necessary for university level work (Hamilton 1991; Sable 1993). On the average, 50% of Aboriginal students drop out before completing high school (Micmac-Maliseet Nation News, March, 1998). Similar statistics can be seen for indigenous cultures throughout the world.[1] This gap is disturbing to indigenous people and educators throughout the Maritimes. The reason for it has not been determined. Explanations suggested by researchers include incompatibility between the pedagogical methods employed by teachers in Aboriginal classrooms (Battiste & Barman 1995; Allen 1995; MacPherson 1987), the paucity of Aboriginal cultural content in the science curriculum (Battiste 1995; Sable 1993,1996), institutional racism, and social structural factors such as issues of class, gender, labour relations and markets, socio-economic inequalities, and so forth (Wotherspoon & Schissel 1998: 35; McKay & Myles 1995). This research illustrates how assumptions by North American educators about what it meant to be educated served to alienate and discount the cultural integrity of a majority of indigenous peoples. It is a testimony to the failed attempts to enculturate and assimilate indigenous peoples under the guise of education.

A review of three important documents produced over the last decade regarding the teaching of science in the United States and Canada reveals that science is a way of knowing, a world view into which we are enculturated from the first day of school. This process of enculturation has to be recognized from the beginning in researching indigenous knowledge systems.[2] Also in these documents is the inherent notion of what it is to be good citizen of these countries. The idea of good citizenship implies values and truths that are legitimised and propagated through the schools, much as Foucault discusses. In my model, these values and truths would be placed in the outermost circle (Figure 8.1). Needless to say, this notion of citizenship would affect the identities of individuals and cultural groups (the inner circles of my model) within the nation/state. These documents substantiate the claims made by indigenous peoples and others that science is a dominant worldview. It is because of its objective pretensions that they question whether a dialogue can be created between indigenous knowledge systems and science as it is taught in U.S. and Canadian schools.

First, in each of the opening statements of these documents, science, mathematics and technology are presented as the cornerstones of human survival in the future in a rapidly changing, competitive, global economy. There are some indisputable facts in these statements, e.g., increasing global interdependence, the need for a sustainable environment, and rapidly changing circumstances, but there is also a disturbing tone particularly in the Benchmarks statement. We are told there will be radical change in the 'terms and circumstances of human existence,' a powerful and frightening statement (AAAS 1993: xi). What are the 'terms' of human existence and who defines these? Next, 'science, mathematics and technology will be at the center of that change' (1993: xi). Consciously or unconsciously, we are being told that these three and not humans are at the centre. They are 'causing, shaping and responding to that change' (1993: xi), like some divine trinity. Consequently, the key element in building a strong future for Canada's young people is science education (Council of Ministers of Education, Canada 1997: 5).

Second, there is a call to equalize access to scientific learning so that all students will reach a basic level of science literacy (1997: 4). The playing field is supposedly being levelled, but the major question for indigenous peoples in both these countries is who is setting the rules? Science, again in its divine role, is the empowering force. 'The potential of science to inform and empower decision-making by individuals, communities, and society as a whole is central to achieving scientific literacy in a democratic society' (1997: 10). Statements like this have led critics, such as science educator, William Cobern, to state, '... the scientific community projects in the public square a pyramid view of epistemology with the natural sciences, of course, occupying the top most position' (Cobern 1999: 6). This pyramid view cannot be taken for granted in a cross-cultural dialogue. Any genuine attempt to bring benefit from international development therefore must allow for other models of knowing or perceiving (Clammer this volume).

Third, along with this comes a stronger emphasis on relating science to the larger context of society and the environment, and promoting the learning of science as being part and parcel of good citizenship. The vision is to provide students the scientific background to understand the many issues, personal and social, that impact upon their lives, and help them develop decision-making skills to deal with the issues they will face as citizens (National Research Council 1996: 107).

Finally, science is seen as a 'unifying force in our society', adding to the 'vitality of the nation' (1996 ix). What is being implied is that science has become so intricately woven into the fabric of our society that science and scientific literacy are unifying forces that will cure alienation and bring all citizens (if they all become scientifically literate) together. Science itself is not being viewed as a possible cause of alienation, but rather as a way to give people a voice and place in society through science education.

Scientific literacy is viewed as essential to a democratic and globally competitive society, whose citizens are empowered (by science) to be active and informed participants in economic growth. The fundamental assumptions or worldview of science education or what science educator, Masaka Ogawa, has termed the 'culture of scientists' (Ogawa 1999), are not open to question by persons educated in other cultural systems. 'From their very first day in school, students should be actively engaged in learning to view the world scientifically' (AAAS 1993: 6). It is this ideology that indigenous people are questioning, and resisting, and it is this ideology that many researchers continue to carry out into the world. It is the model for how we study others. This is a demonstration of the process of enculturation within education.

Henry Giroux (1983), in his call for critical discourse theory in education, addresses the rationality that lurks behind this type of citizenship education, noting that '...it has been largely influenced... by the culture of positivism, with its underlying technocratic rationality' and is 'rooted in a myriad of political and normative issues' (Giroux 1983: 193). Even those educators or researchers who are sensitive to different world views and constructions of reality can remain ignorant of the larger ideological context in which any project, educational or otherwise, occurs (1983: 176–186). This is what is commonly referred to as the 'hidden curriculum' in education (1983:198), but can also be thought of as the hidden agenda of development projects. Clammer views this as the process by which indigenous knowledge is being 'harnessed to a project or a grand narrative-development...' (Clammer 2002).

Dr. Gregory Cajete, a Tewa Pueblo educator, describes the path of knowing as an inward journey versus what he perceives as an outward journey taught in the western educational system. He does not discount the contribution of objectivist research but notes its limitations and lack of attention to '...the affective elements – the subjective experience and observations, the communal relationships, the artistic and mythical dimensions, the ritual and ceremony, the sacred ecology, the psychological and spiritual orientations that have characterized and formed Indigenous education since time immemorial' (Cajete 1994: 20). The 'traditional' in traditional ecological knowledge is not just about what indigenous peoples know but why they care I extend Giroux's intentions, giving his concept for 'emancipatory rationality' an anthropological application, along with Smith's cautionary note regarding the potential for arrogance no matter what the theoretical approach. In Giroux's work, he is looking to develop a critical theory of education. It is intended to help us 'see educational practices as historical and social products...' (Giroux

1983: 195). This is in keeping with Holland and Levinson's concept of the cultural production of the educated person. Again it is a dialectical process to locate the problematic inherent within any knowledge system, and a method to reveal contradictions. I use this self-critical thinking as a way to realize our own cultural blind spots.

It is not a question of proving one knowledge system as superior to another or creating 'cut-and-dried binary discriminations between "us" and "them"' (Sillitoe this volume). In fact, as the following interviews indicate, it is possible for people to maintain multiple rationales and move between them as foreground and background gestalts depending on the context in which they find themselves. By applying some of the concepts of dialectical process, we can begin to bring the foregrounds and backgrounds into a dialogue to enrich all. But it requires that we must follow Giroux's advice that knowledge must be 'stripped of its objective pretensions' when dealing with another cultural community.

Creating a dialogue between cultures

From this critical perspective, we can begin the exploration of how people from any particular culture perceive, cognise, experience and express reality and how that is passed on through time. Vine Deloria's suggestions regarding how we might compare a scientific way of collecting and interpreting data with that of indigenous peoples is helpful.

> The first question to be asked prior to a comparison of the results is how the two groups gathered information. What makes the data that each group considers in making decisions about the natural world reliable either in their own eyes or in the view of impartial observers? The initial approach to experience, the process by which we identify data as important or irrelevant determines how we choose to connect experiences with facts. We formulate the subsequent patterns of arrangement and interpretation, which we impose on experience after this initial identification (Deloria 1992: 3).

In my own research on these questions, I arranged to observe a Grade 7/8 class within a Mi'kmaw First Nations school. The school was formed to counteract what some members of the community perceived to racist attitudes in the off reserve, provincial school, and a means to reclaim their own culture. Permission was given by all the parents of the children in the class, as well as from the students, the teacher, the principal and chief of the reserve. I arranged a meeting with interested parents prior to beginning my visits to answer any questions they had about the motives and goals of my request. I visited the school over a four-month period, one morning a week, and taped, observed, and took notes within the classroom. Twenty-five hours of videotaping was conducted within the classroom through various subject periods and during field trips, and general discussions were held with the staff and teachers at the school. Two special field trips were arranged for the children led by the head of a forest management project within the community. This person was researching and integrating into his work the traditional ways his ancestors used the land. He also had a university education in forest management.

The field trips were intended to see if a dialogue between traditional knowledge of the Mi'kmaw and current scientific teaching could occur.

The class consisted of eleven children, nine of who granted permission to be interviewed. One dropped out of this commitment, leaving a total of eight students who were interviewed. These students chose to be interviewed in pairs. The interviews were conducted in a library-resource room within the school, and were semi-structured, lasting 45 minutes to an hour. The questions ranged over a large area of topics concerning their school and community life, their attitude toward and understanding of science, their knowledge of their own traditions, and well a their own identity as Mi'kmaq.

One of the many questions I asked students in interviews was how they defined science. From their answers, it was clear they appeared to have very little connection with science. Science is about 'things' and how things work, an operational explanation of the world, but that is not quite the same as truth, or what is personally meaningful. In fact, in my observations of these First Nations students, science was not even practical for the most part because it was so inaccessible to them conceptually. The language of science itself was like a foreign language. Experientially, science was seeing things happen, e.g., a test tube explode, or a chemical reaction. It was directly observed action events. When asked what they liked most in science they responded by choosing chemistry because you see things happen. Only one student out of eight mentioned physics as interesting. This is not unique to indigenous children but can be seen in science education research throughout the world (Aikenhead and Jegede 1999).

To establish a link between their western view and their own cultural heritage I asked them how they tell if something is true. The following is a conversation I had with two of the students. It illustrates that how we determine something to be true is culturally rooted and how what we believe to be true shapes our whole relationship with the world.

TS.   Can you tell me how you tell if something is true?

S1.   When it's right there. Or when you know it's there or something.

TS.   What about when you read things in books? Is there something you consider to be...

S2.   Yeah. If you think it's the truth, and you believe it's the truth, then it probably is. But if you have second thoughts then you don't...if you don't like say that it's right 100%, then it probably ain't.

TS.   So what if someone came running up to you and said, 'I saw a ghost' or something, or an alien was there'.

S1.   If they were crying I would believe them.

S2.   Yeah. If you knew this person real good, and you knew that they wouldn't lie to you, then you would have to believe it.

TS.   But you said something is true if you can see it...

S1.   If you were there and saw an alien, you would know it's true but if you tried to tell somebody else, nobody would believe you. They would just think you were trying to be on the news or something.

TS.   So, it takes a lot of courage, I guess.

S1.   Yeah. My grandmother tells me all about these old Indian legends like about the story of the *pukulatmu'jk* and all that stuff. These little men that live by the river.

S2.   We have *pukulatmu'jk* caves down there.

TS.   Do you?

S2.   Yeah.

S1.   Yeah there's little caves in the side of the river where there's like a cliff, not very high, not even as tall as this school eh? Just about a few feet. Maybe like 15 or 20 feet?

S2.   There's a lot of stories.

S1.   Yeah, and like...

S2.   *Pukulatmu'jk* is definitely the biggest.

S1.   Yeah, there's a lot of stories about those. My grandmother used to always tell me them.

S2.   We went inside one like two years ago when we went to get clay down the river.

S1.   Oh yeah. We went inside one of those caves.

S2.   It was small.

S1.   We were like....

S2.   It was like, 'Stay here. We might hear footsteps.'

S1.   Yeah, it was like a class and we were all like protecting each other. But it was big enough to walk in.

S2.   'Three at a time, three at a time.' (They are sort of laughing as they are talking.)

S1.   And it was like go right through. It was covered and then there is like another spot at the end and then there's tunnels going through.

S2   But just in case they are real we don't want to be intruders. Like now, when you're going up to (place name), you see the signs that say deer crossing and the only reason that they're out there is because the people are cutting in the woods and stuff and they're destroying their homes. So, when we went down, we were just going to look in there out of curiosity. We weren't going to destroy it or...

TS.   You're talking about the *pukulatmu'jk* now? You mean if you just go and look you won't be hurting them or anything. Is that what you're saying?

S1.   Some people say about the *pukulatmu'jk*...

S2.   ... that they're evil.

S1.   Yeah, they're evil cause they ran away from crosses or they knocked them all down one time when the settlers or something came.

TS.   You mean like Christian crosses? Is that what you're saying?

S1.   Yeah. And some say they're real nice and they'll help you out of the woods and some say that if you believe in them...

S2.   ...that they sit on the side of your bed and they laugh at you while you're sleeping.

TS.   Is that good?

S2.   That's not good.

TS.   If you believe in them that they'll sit inside your bed and laugh at you?

S2. No, if you believe in them, then they are real, or you might be able to see one but if you think it's a myth, and you think that it's just like just something to get your kids scared then it's not real.

TS. What do you think (to S1)?

S1. I think both.

TS. Real and not real?

S1. Yeah.

S2. Yeah.

TS. That's interesting. Can you say more? How you can say things are real and not real at the same time?

S1. Like when little kids are scared of ghosts and you know that they are not real but some people that believe in them...

This interview reveals the fluid construction of reality. In this conversation aliens are said to be unreal or imaginary, but real if a trustworthy person, not in it for personal gain or publicity sees them. This was reflected in another conversation in which the student said truth was based on whether you believed the person telling you. There was some degree of social consensus that *pukulatmu'jk*, or little people, exist in a place located within the community. If one believes in them, they are real.

What makes this culturally distinct from a western view is that some elders would support their views, and social consensus, particularly supported by elders, is how truth is determined. It is the elders in the community that represent the source of cultural continuity to the extent that they are still trusted. When children talk about little people or other types of beings in the world as possibly real, it shows their notions of reality can persevere even though our scientific notions have been embedded in their education for over a century.

No doubt people not of Aboriginal descent have similar concepts or parallel versions of reality as the ones voiced by these students, e.g., ghosts in a graveyard. (Aikenhead and Jegede 1999). However, it is what is socially or culturally accepted as reality that are really at issue here, as well as how these students come to decide what is true for themselves. These views are embedded in the myths (discourses) of that culture, whether they seem to appear as superstitious tales or as concurrent with scientific theories. Truth then becomes a matter of social consensus. It is no wonder that some indigenous peoples may be loath to have their whole cultural way of relating to the world undermined by the ideological exclusion implied in science.

Anthropologist Irving Hallowell understood the power of the narratives that people use to contextualise these truths, noting '...what people choose to talk about is always important for our understanding of them, and the narratives they choose to transmit from generation to generation and listen to over and over again can hardly be considered unimportant...' (Smith 1995: 19). This is what Foucault referred to as the 'production, accumulation, circulation and functioning of a discourse' (Foucault 1980: 93) that is the foundation for a dialogue. It is the power of the truths within this discourse that affect people's actions and are the basis for decision-making. Cultural beliefs merge with practical issues, such as caring for beings that inhabit the environment. The implication is not so much that development agents or scientists must sincerely join with another culture's belief systems. The point is rather to demonstrate respect for the entire cultural matrix that supported this culture within a particular landscape for thousands of years.

This story is also an illustration of cultural beliefs that have endured through 400 years of contact, which is the 'traditional' aspect of 'traditional ecological knowledge', and one example of how these children relate to their world. The caves within their community are not just made of inanimate rocks and soil, but are to them the habitat of persons who required respect and awareness of their presence. Even if the children do not choose to believe in them unconditionally, *pukulatmu'jk* are still *possibly* real. Mixed in with that possibility, the students understand that disturbing these caves is similar to woodcutters destroying the habitat of deer, something they study in their science textbooks under the heading 'Diversity of Life'. The concept that one should respect life, preserve habitats and diversity of life, and understand the consequences of disruptive behaviour is an inherent truth in both the science lesson and the stories of *pukulatmu'jk*. Here is a concrete basis for a dialogue, a common ground, a cross-cultural insight. It is a possible intersection of understanding described by Rabinow earlier in this chapter.

To bring this dialogue into a more concrete level for curriculum development, I attempted to see if the students could make a connection between traditional knowledge and the science they were being taught. The students interviewed were taught science using the same textbooks as all children throughout the provincial school system. These students also had a Mi'kmaw language and indigenous culture class twice a week. During one of these culture classes, a film was shown of a Cree man making a canoe. In it, he measured wood, boiled sap to just the right temperature, used a variety of materials to shape and build the canoe, etc. During an interview with two students that same afternoon, I attempted to see if they could connect what they had learned about the land from their parents or grandparents, as well as the many legends they had been told, with scientific knowledge. In this interview, I used the canoe-making film as my entry point. The children had just been studying physical and chemical change in their science classes.

TS.   The other day you were talking about physical change right? That same day, you had that canoe making film. Remember?

S1.   He knew how to make the wood bend...

TS.   And also the sap, blending the sap and the fat and heating it to the right temperature. You know, it was all about chemical change. Would that be interesting to you? That kind of tying that....

S2.   Yeah, like Indian people make those kinds of boats. And it's kind of like science too because like when you do something with chemicals, you have to make sure it's the right amount, the right temperature, and the right kind, and just like how they were making that sap and the fat and that kind of stuff.

S1.   And like dying quills and stuff and they know how to get them, they know how to cut them so they won't get hurt, they know how to dye them, they know how to put them together.

TS.   Do you think that would be interesting? I mean, I'm not trying to say you shouldn't learn science as it is now in the books, but when you look at it, there was all this knowledge people had. What do you think?

S1.   That would be cool cause it would be like part of their culture.
TS.   I'm not saying don't learn science but rather create a dialogue.
S1.   They're both important.

This research attempts to establish a dialogue, not a debate, between knowledge systems. It seeks to find a ground of mutual meaning, where knowledge can be looked at from different cultural perspectives. These two students made the connection between their ancestors' technical knowledge about physical change, and a western scientific view. They could appreciate the integrity of both. Once the dialogue is opened and the students make the connection between their ancestral knowledge and their contemporary education, they can begin to look further at the 'webs of significance' for themselves. They can begin to look at the assumptions people make about their world, and the truths that informed the gathering and application of knowledge, much as Deloria suggested. These two different frames of reference do not have to be equated nor conflated but rather brought together in a dialectical process that attempts to address the shifting frames of meaning in the contemporary world.

Conclusion

This chapter is about how we could enter any community with proposals for change and 'development' and create a meaningful dialogue. To establish a dialogue first requires a process of self-reflection on our part as the researchers or development agents, to understand the models of thinking, ideologies that fuel those models, and motivations and agendas that we bring to any proposed project. This is the process of what Giroux (1983) termed 'stripping bare objective pretences of knowledge'. This process creates a more open and productive ground for working with others in a dialogue.

Three national documents concerned with science education in the U.S.A. and Canada were critiqued as an example of how education enculturates students into specific ways of thinking and determining the nature of reality as scientifically defined. This way of thinking is deemed critical for students to become productive citizens within a democratic, globally competitive society. They provide an example of the cultural production of knowledge within these developed countries, as well as point to the ideologies researchers carry with them into other cultures. Understanding this is primary to beginning any dialogue within other cultural settings.

The foundation for a dialogue is understanding the shared truths that inform what it means to be knowledgeable within any culture. I have supported Levinson and Holland's theory that the educational systems of any culture, in whatever form they exist, are where the cultural production of the educated person is negotiated and defined. Understanding what it means to be educated is understanding what knowledge is deemed culturally relevant to the mental, physical and spiritual survival of people. It is within these systems that truths are negotiated and knowledge determined as meaningful. It is also within these systems that the emergent nature of culture is evident, and where the various local/state/national discourses come into play. The educational contexts are where personal and cultural identities are formed.

The student interviews point to the potential for a dialogue as a means to create a mutual ground for understanding and meaning between different cultural perspectives. The dialectical process is a journey to discover intersections of understanding on which further dialogue can be built. That process will inevitably lead to learning on both sides.

An ongoing dialogue is the means to create a common ground of exploration and creation of meaning between different knowledge systems. If properly constructed, this dialogue will ensure cultural continuity for the culture in which change is being initiated. For this dialogue to occur, it is our challenge to understand the cultural contexts in which we all exist, and how deeply they inform our concepts of truth and, in turn, what it means to be knowledgeable. Without these underlying truths, knowledge becomes meaningless because it has no personal or social context; it can have a negative social impact and discount cultural continuity.

The challenge is to understand how people negotiate and determine truths necessary for cultural continuity within shifting realities of emerging cultures. Knowledge within any culture is about relationships whether human or not, and it is within those relationships that one finds identity and power. If development projects wish to succeed, it is necessary to work within these relationships versus creating conflicting versions of truths and realities, which breed resistance and failure.

Acknowledgments

I would like to very specially thank Stephen Ginnish for the field trips he organized and led, and for his consistent support and feedback throughout this project. This gratitude also extends to the wonderful and tolerant teacher of the Grade 7/8 class I observed, the Grade 7/8 students, the principal, teachers and staff of the school, the chief and band council, and Stephen Ginnish's wife for their cooperation and support.

I would also like to express deep gratitude to Dr. Colin Dodds, Dr. Mike Larson, and Dr. Paul Erickson of Saint Mary's University in Halifax, Nova Scotia, Canada for their approval of the funding that made my attendance at the Association of Social Anthropologists 2000 conference in London possible. Finally, I would like to thank my husband, David Sable, for his insights and feedback.

Notes

1   See, for instance, 'Progress Towards Closing Social and Economic Gaps Between Maori and Non-Maori: A Report to the Minister of Maori Affairs.' Te Puni Kokiri, Ministry of Maori Development, New Zealand, 1998.
2   These three national documents are: Benchmarks for Scientific Literacy, 1993, published by the American Association for the Advancement of Science (AAAS), National Science Education Standards, 1996, published by the National Research Council, and Common Frameworks for Scientific Literacy, 1997, produced by the Council of Ministers of Education, Canada.

References

Agrawal, A. & C. Gibson. 1999. Enchantment and disenchantment: The role of community in natural resource conservation. *World Development* 27(4): 629–649.

Aikenhead, G. & O.J. Jegede.1999. Cross-cultural science education: A cognitive explanation of a cultural phenomenon. *Journal of Research in Science Teaching* (36): 269–287.

Allen, N. 1995. *'Voices from the bridge' – Kickapoo Indian Students and Science Education: A Worldview Comparison*. Unpublished Ph.D. Dissertation. University of Texas.

American Association for the Advancement of Science.1993. *Benchmarks forscientific literacy*. New York, Oxford: Oxford University Press.

Arkwright-Alivisatos, D. 1998. *Preparing for the next millenium: A profile of New Brunswick Aboriginal High School students*. Report in progress for the New Brunswick Department of Education, Fredericton, New Brunswick.

Battiste, M. 1995 Introduction. In *First nations education in Canada: The circle unfolds*. (eds.) M. Battiste & J. Barman. Vancouver: University of British Columbia Press. vii-xx.

Battiste, M. & Barman, J. (eds.) 1995. *First nations education in Canada: The circle unfolds*. Vancouver: University of British Columbia Press.

Bauman, R. 1977. *Verbal arts as performance*. Prospect Heights, Illinois: Waveland Press.

Cajete, G. 1994. *Look to the mountain: An ecology of indigenous education*. Durango, Colorado: Kivaki Press.

Chatwin, B. 1987. *The songlines*. New York: Viking Penguin.

Clammer, J. 2002. Beyond the cognitive paradigm: Majority knowledges and local discourses. In *'Participating In Development': Approaches to Indigenous Knowledge*. (eds.) P. Sillitoe, A. Bicker & J. Pottier. London: Routledge (ASA Monograph Series No. 39): 43-63.

Cobern, W. 1999. *The cultural nature of the concept 'scientific worldview'*. Paper presented at the National Association of Research in Science Teaching, Boston, Massachusetts.

Council of Ministers on Education, Canada. 1997. *Common framework of science learning outcomes: Pan-Canadian protocol for collaboration on school curriculum*. Council of Ministers on Education, Canada, Toronto, Ontario.

Deloria, V. 1992. Ethnoscience and Indian Realities. *Winds of Change* Summer. 12–18.

Edwards, M. 1999. NGO performance – What breeds success? New evidence from South Asia. *World Development* 27(2): 363–374.

Ermine, W. 1995. Aboriginal Epistemology. In *First Nations education in Canada: The circle unfolds*. (eds.) M. Battiste & J. Barman (pp.). Vancouver: University of British Columbia Press. 101–112.

Foucault, M. 1972/1977. *Power, knowledge: Selected interviews & other writings, 1972–1977*. C. Gordon. (ed.) New York: Pantheon Books.

Geertz, C. 1975. *The interpretation of cultures: Selected essays of Clifford Geertz*. London: Hutchinson & Co.

Giroux, H. A. 1983. *Theory and resistance in education: A pedagogy for the opposition*. South Hadley, Massachusetts: Bergin & Garvey Publishers.

Hamilton, W.D. 1991. *Closing the gap*. Report prepared for the New Brunswick Department of Education, Fredericton, New Brunswick, Canada.

Haraway, D. 1989. *Primate visions: Gender, race, and nature in the world of modern science*. New York, London: Routledge.

Hess, D.L. 1995. *Science and technology in a multicultural world: The cultural politics of facts and artifacts*. New York: Columbia University Press.

Kawagley, A.O. 1995. *A Yupiaq worldview: A pathway to ecology and spirit*. Prospect Heights Illinois: Waveland Press.

Knorr-Cetina, K. 1981. *The Manufacture of knowledge: An essay on the constructivist nature of science.* Oxford, New York: Pergamon Press.

Levinson, B.A., Foley, D.E. & D.C. Holland. (eds.) 1996. *The Cultural production of the educated person: Critical ethnographies of schooling and local practice.* Albany, N.Y.: State University of New York Press.

Levinson, B.A. & D.E. Foley. 1996. The cultural production of the educated person: An introduction. In *The Cultural Production of the Educated Person: Critical Ethnographies of Schooling and Local Practice.* (eds.) B. A. Levinson, D.E. Foley, & D C. Holland. Albany, N.Y.: State University of New York Press. 1–54.

Mackay, R. & L. Myles. 1995. A major challenge for the education system: Aboriginal retention and dropout. In *First Nations education in Canada: The circle unfolds.* (eds.) M. Battiste & J. Barman. Vancouver: University of British Columbia Press. 157–178.

McPherson, J. 1987. Norman. In *For the Learning of Mathematics* (7) June, 2: 24–26.

Micmac-Maliseet News. 1998. 1996 Census: Aboriginal data. In *Micmac-Maliseet Nation News.* March, 1998.

National Research Council. 1996. *National science education standards.* Washington, D.C.: National Academy Press.

Ogawa, M. 1999. *Science as the culture of scientists: How to cope with scientism?* Paper Presented at the National Association of Research in Science Teaching, Boston, Massachusetts.

Rosaldo, R. 1989. *Culture & truth: The remaking of social analysis.* Boston: Beacon Press.

Sable, T. 1993. *Development of an earth science curriculum for Mi'kmaw children.* Report to the Department of Energy, Mines and Resources, Natural Resources, Canada.

Sable, T. 1996. *Another look in the mirror: Research into the foundation for developing an alternative science curriculum for Mi'kmaw children.* Unpublished M.A. Thesis. Saint Mary's University, Halifax, Nova Scotia, Canada.

Schneider, G. 1998. Reinventing identities: Redefining cultural concepts in the struggle between villagers in Munda Roviana Laggon, New Georgia Island, Solomon Islands, for the control of land. In *Pacific answers to western hegemony: Cultural practices in identity construction).* (ed.) J.Wassmann. New York: Berg. 191–211.

Schultz, E. A. & R. H. Lavenda. 1995. *Anthropology: A perspective of the human condition.* Mountain View, California: Mayfield Publishing Company.

Scott, J. 1990. *Domination and the art of resistance: Hidden transcripts.* New Haven and London: Yale University Press.

Smith, L. T. 1999. *Decolonizing methodologies: Research and indigenous peoples.* London: Zed Books and Dunedin, New Zealand: University of Otago Press.

Smith, T.S. 1995. *The island of the Anishnaabeg: Thunderers and water monsters in the traditional Objibwe life-world.* Moscow, Idaho: University of Idaho Press.

Stephenson, N. A. 1998. Contrasting transcripts: Constructing images and identities in mediations among the Wam people of Papua New Guinea. In *Pacific answers to western hegemony: Cultural practices in identity construction).* (ed.) J.Wassmann. New York: Berg. 143–168.

Te Puni Kokiri Ministry of Maori Development. 1998. *Progress towards closing social and economic gaps between Maori and non-Maori.* A Report to the Minister of Maori Affairs, New Zealand.

Trujillo, A.1996. In search of Aztlan: Movimiento ideology and the creation of a Chicano worldview through schooling. *The Cultural production of the educated person: Critical ethnographies of schooling and local practice.* (eds.) B.A. Levinson, D.E. Foley & D.C. Holland. Albany, N.Y.: State University of New York Press. 119–149.

Wotherspoon, T. & B. Schissel.1998. *Marginalization, decolonization and voice: Prospects for aboriginal education in Canada.* Discussion Paper. Pan Canadian Education Research Agenda, Council of Ministers of Education, Canada.

Chapter 9

The Re-emergence of Traditional Medicine and Health Care in Post-Colonial India and National Identity

Subhadra Mitra Channa

The concept indigenous knowledge has emerged out of post-modern discourse on development, which has engaged anthropologists among others and informs the critique of earlier structuralist and modernist approaches which had privileged Western science and capitalism as the ultimate arbiters of development. In the eighties, approaches such as 'participative development' (Pottier 1993) came in to vogue which recognized knowledge systems other than western ones and also complied with open market theory (Sillitoe 1998). However these have continued to treat the indigenous differently assuming of the existence of a privileged Western category often including the anthropologist (Moore 1994: 5). Even at the Association of Social Anthropologists, 2000 conference it was assumed by some that the indigenous was something the west was not. In this sense the superior and the knowing voice was always that of the anthropologist either from the west or trained in the west who was in a position to study and construct a discourse on the indigenous. One may take as an example this single sentence. The cross-cultural study of *their* knowledge may advance *our* scientific understanding of natural processes by challenging *our* concepts and models (emphasis mine) (Sillitoe 1998: 227).[1] The fact that highly systematized systems of knowledge such as the Indian Ayurveda are classified as 'indigenous' (Hobart 1993: 4) indicates the western bias in its conceptualization. For Ayurveda is a very ancient and systematic knowledge supported by scholarly written texts (Alter 1999) Even if we accept Veronica Strang's (n.d.) identification of indigenous with the local or a 'landscape' we must not overlook the colonial history behind the global expansion of western science from its European origin. Power more than efficacy can be seen as the success story of the globalization of western science. Vitebasky (1993: 106) has written that expansion of any knowledge beyond the frontiers of its origin is necessarily by a process where it must 'annihilate, degrade or subsume other ways of knowing'.

The power equation in defining what is to be legitimized as knowledge has been recognized by many writers (Chatterjee 1989: 623; Hobart 1993: 16; Cheater 1995: 122). The colonial powers established moral justification of their rule by labelling the colonized as ignorant and in need of being 'educated' and 'ruled'. As Hobart (1993) points out it entailed the devaluation of their systems of knowledge, and as Chatterjee points out, of their traditions as well. Furthermore anthropological voices are more often than not presumed to be western as recognized by many (Moore 1994: 5; Ferrades 1998: 259; Gledhill 1994: 4).

In this paper I take exception to this privileging of western voices in the discourse on the indigenous. The Indian anthropologist Beteille (1998: 190) criticizes the essentialist view of the term 'indigenous'. Does indigenous always refer to the 'other'?[2] Even if we take the essentially Western perspective on indigenous as 'that which is not western', which in itself is a problematic concept (Moore 1994: 130) it may refer not to the 'other' but to 'mine', through a simple reversal of the position of the speaker. For a non European, the indigenous would refer to 'mine'. This reversal of identities throws a different light on the indigenous especially in a political and economic frame of reference. I am not an anthropologist for whom the indigenous is the 'other'. For me it is what is 'mine' because I belong to a culture dubbed indigenous by, for me, the 'other'. But is this the only meaning and significance of indigenous knowledge? Does the concept of indigenous knowledge have no existence or significance outside of the discourse of development? This paper considers the political construction of the concept of indigenous knowledge and its use as a tool to construct an effective critique of colonialism by the colonized. It understands knowledge as power (Foucault 1972; Bourdieu 1977), not simply knowledge as development. It focuses on the construction of indigenous knowledge from a non-western point of view. In this perspective indigenous knowledge is not only seen as 'mine' in contradiction to 'theirs' but it also indicates different levels of understanding of indigenous knowledge, which is based not upon a western and non-western dichotomy but the divisions internal to the 'non-western'. The indigenous is not a homogenous category but comprises many layers of meaning informed by both relative power and situation.

Taking India not simply as a case study but as a lived reality for my self, I shall explore the different ways indigenous knowledge is defined and used, for different goals by different sections/class of people. The concept may be borrowed from the West but understood and interpreted both with reference to its local and global use. I try to show that different sections of Indian society are constructing and interpreting the concept of indigenous knowledge to suit their political goals.

I consider several sections of Indian society, the educated elite and the ordinary middle class, the rural poor, the NGOs and the political power holders. Each of them has a different perspective on traditional or non-western systems of health care. This chapter focuses on this aspect of indigenous knowledge, the indigenous healing systems. Further the non-western, or what for simplicity's sake we may continue to call indigenous system of medicine, refers to not one but several different systems each of them being accorded different political and economic status. As (Pottier n.d.) points out 'local knowledge is not homogenous ... some perspectives dominate while still others are marginalized'.

Theoretically, I take indigenous knowledge as a 'thing in itself' and examine its political and cultural significance in historical and class relative contexts. Not only is indigenous knowledge looked upon and used differently for different ends by different people, it also has a history. I also endorse the view of Cohen (1993) that indigenous knowledge is constituted in social relations and in spite of being internally diverse may serve as a 'cultural totem' to define community identity.

How is indigenous knowledge understood in the Indian context?

In Hindi[3] the nearest equivalent of the term indigenous is 'Desi', and thus what may be understood as indigenous system of medicine is 'Desi Dawai', which means literally indigenous medicine. Reflecting on the meaning of 'Desi' or indigenous here problematizes the concept. In common linguistic usage 'Desi' is opposed to the term 'Bidesi', which means foreign or not of local origin. But in relation to indigenous medicine, the term is opposed not to 'Bidesi' but 'Angrezi', which literally means English. Moreover medicine referred to as 'Desi' is not understood as of local origin but that which is not 'Angrezi'. It makes clear a power differential in the definition. Thus English is also to be read as colonizer or ruler. That the term 'Angrezi' is not coterminous with Western can be seen by the various systems not included under it. Thus the term 'Desi Dawai' refers to a variety of medical systems, Ayurvedic, 'Hakimi' or 'Unani', Homeopathy, 'Jhaar Phoonk' or magical cures and home remedies. If we take indigenous to mean 'of a local origin' then neither Unani nor Homeopathic falls into this category. Unani or Hakimi refers to a system of medicine supposedly originating from Iran (Unan) and brought into India by the Muslim invaders and by and large practiced by Muslim healers. Homeopathy originated in Germany and is practiced by anybody. Ayurvedic refers to higher systems of Hindu learning as embodied in the Vedas and is usually associated with a Hindu and Brahmin practitioner. The magical systems of cure are local and part of oral traditions and usually practiced by low castes. The home remedies are the forte of women and are the grandmother's cures found in most families, but some women may be considered expert practitioners. Some people may take Homeopathy as a separate system not explicitly included under 'Desi' but it is never understood as Angrezi which is reserved for what may alternatively be called allopathic[4] medicine but always called 'Angrezi' in the Hindi usage.

Apart from its colonial associations with the British for most Indians the 'Angrezi' system of medicine is dissociated from their worldview regarding the body and its state of well being. Thus what is considered 'Desi' or 'Angrezi' may well refer not to the source or origin of the medicine nor to its situation but to the compatibility between it and local ways of thinking. Thus 'Desi' may refer to 'What is like ours' as well as 'What is ours'.

Cognitive aspects of health and cures

a)   The Karmic principle of disease and the doctor as a supernatural interventionist

A large number of people in India from all walks of life, view illness of the body as a result of bad deeds done sometime in the past (period unspecified) or to one's fate or kismet. Although they derive from two different world views one Hindu and the other Muslim, the two are not incompatible at the practical level. People may, with equal ease, refer either to bad karma or bad kismet, although the implications are entirely different (karma refers to a cause and effect relationship and kismet to uncontrolled destiny). Many people when ill lament their 'own deeds' which have led them to such a state or to their unlucky stars. 'I am suffering because of my own karma' or' what bad deeds did I do to suffer this ', 'my kismet is bad 'are frequently heard refrains from patients. This attitude also colours the way the medical practitioners are viewed. The doctor is seen not simply as any other specialist but one with mystical powers to fight the results of one's 'karma' or kismet. Medical practitioners of every kind are clothed in this aura of supernaturalism. This is expressed in the way a patient addresses a doctor and also in the overall reverence shown towards him or her. A doctor is always addressed with an honorific like Saheb, Babu etc. affixed to the term doctor. Thus a doctor in Northern India is always addressed as Doctor Saheb even by most of the urban elite. Moreover people turn to a doctor for more than just medical assistance. The concept of medical care encompasses much more than simply medicines and consequently the doctor is rewarded with more than just fees. Gifts are lavished, women knit sweaters, and men make personalized gifts. Often this may mean something made with one's own hands.

The doctor's success also depends on matching such an attitude. Most patients are not impressed by the brusque and matter of fact attitude of the modern medical practitioner. What in the West is seen as professionalism would be frowned upon in India. Western medical science views the human body as a passive object in relation to the medical practitioner (Lyon and Barbalet 1994: 48; Jordanova: 1980). A spiritual and unified conceptualization of the body is more common than the objective one.

This separation of the body from the human person is not found in cultures other than those dominated by western science. The secular view of body as an object has conditioned the concept of illness in western medical science as a purely objective condition to be treated as such.

b)   What is an illness?

There is wide variation in the way illness is perceived and needing medical attention. While working in a village in Rajasthan I was told by most people that snake bite was not a condition for which one went to a medical practitioner, of whatever variety, one went to a Bhopa (a local equivalent of a shaman). This was all very well for people whom I thought were not educated enough to know the real effects of snakebite and its ready treatment with medicine. But then I had along discussion with an educated man in the village who was a postgraduate, and who lamented the ignorance of the so called uneducated people in the village and their ignorance of modern medical practices. 'We go to the hospital for everything' he ended up quite emphatically.

'Even for a snake bite?' I ventured to ask. At this the man appeared surprised at what he thought was my ignorance. 'No, no, we go to the hospital for all illnesses, snake bite is not an illness. For that we go to the Bhopa'.

Not only the condition of the body, but whose body and who is ill also determine the nature of the practitioner resorted to. For example, most people in the hill regions of Garhwal do not take small children to medical practitioners even though they may turn to them for most adult diseases. The reason being, small children are considered especially vulnerable to the supernatural and hence it is more likely that the illness is caused by a supernatural agency. It is thought to be better to take them to an exorcist or shaman first than to waste time in going to doctors. In fact most persons go to a medical practitioner in these regions only when they are sure that the condition they are suffering from is not due to supernatural causes. There are very many conditions which are classified in modern medical systems as meriting medical attention but which Indians regard quite differently. I worked for sometime as an assistant to a psychiatrist, in a nursing home, situated in the heart of urban Delhi. It was a revelation how many middle class, fairly well educated persons thought of mental illness as having a supernatural cause. One woman whose son was schizophrenic asked me, in confidence. These doctors here are all male, they do not understand, but you tell me you are a mother, are you sure my son is not ill because some one has done 'jadu tona' (black magic) on him?' It is due to such deep rooted belief systems that many turn to shamanistic practitioners as shown by Kakar in his 'Shamans, Mystics & Doctors'(1982).

I also worked as the social anthropological advisor to an Overseas Development Agency funded project in a Central Government run hospital in Delhi, The Lok Nayak Jaiprakash Narain Hospital located between Delhi Gate and Turkman Gate near the densely populated walled city of Old Delhi. The aim of the project was to find out why women who were living close to the hospital, that is in the slums that lined the hospital walls, never came to the hospital for treatment in spite of the free and excellent medical care. Investigation revealed that many women were suffering from reproductive tract diseases. The major symptom was a white discharge from the vagina accompanied by lower back pains. They thought that the base of their spinal cord was dissolving and coming out from the vagina as a white discharge, which explained the pain in the lower back. They considered it a disease of the bone and not an infection at all. The local practitioners to whom these women were going gave them medicines that would strengthen their bones and give them 'takat' (energy). They were also advised to take strengthening foods like bone soup and a diet rich in proteins. The treatment increased their strength, they often felt better and many were able to withstand the infection. When they came to the hospital however the first thing the doctors did was put them on a table and subject them to an internal examination which they found both embarrassing and very painful. Also given their understanding of their condition as 'melting bones' they thought the physical examination also very threatening. Moreover the doctors at the hospital were curt and scolded them sharply if they screamed or showed pain. This put them off and few of them ever returned. The experience of one woman was enough to put off many others.

c)   The perception of the body as a moral entity

Indians regard the body as a moral entity and not simply as physical. For Hindus it is the abode of the divinity a spark of which exists within all humans. The western medical system recognizes only the physical existence of the body. The treatment meted out to the body treats it as an 'object'. The practices of western medical science such as 'examination', 'probing', 'insertion' and 'injection' do violence to belief in the human body as the sacred repository of the soul, and also assault such values as purity and honour to be found in both Islamic and Hindu traditions. Most persons in India feel revulsion at the body being subjected to such indignities. Hindus believe that the sacred resides in each human being. The body is thus a temple. In Eastern India, if a person accidentally touches another person's body with his or her foot, s/he will immediately bend down and touch that person's feet as a gesture of apology and atonement. Many of the people I knew who hold traditional ideas avoid injections or even' blood tests'; so much a part of the repertoire of western medical science, not because of the pain, but because of the indignity of such probing and body penetration. Even the younger and educated, especially women, 'simply hate to go' to a practitioner who would prod them and subject them to 'internal examinations'. A major attraction of homeopaths for many educated and well-informed persons is that they provide a respectable alternative to the 'allopathic' doctor who is always 'probing and prodding'.

The homeopathic doctors are holders of degrees and are recognized as registered medical practitioners. They never subject you to a physical examination although they subject you to a long series of questions. But this verbal probing is far more attractive to an Indian clientele as they feel that the doctor is showing interest in them as persons and not simply probing them as 'soul less objects'. The homeopathic doctor asks questions that an allopathic doctor never asks. Such as, 'How the patient feels?' (In an abstract sense) 'What are the patient's moods like?' 'Does a person feel hot or cold?' 'Does one get bad dreams?' These questions make a person feel human and not just a body, objectified without a soul and emotions.

A patient rarely views the illness as a purely physical condition. It is also a moral condition, one of disharmony, a realization of one's sins. A person in such a state is also given to moods and psychological problems of brooding and depression. The approach of the traditional medical practitioners such as the Vaids and the Hakims, as well as the plethora of folk level practitioners, is usually directed towards the mental rather than the physical condition of the patient. Their success depends on targeting many of the psycho -somatic causes of diseases (Kakar 1982, 1986).

Another positive point with traditional medical practitioners is the way in which they examine patients and their mode of diagnosis, which cause least trauma to the patient. The patient is always examined sitting up and facing the doctor. In most rural areas the doctor sits on the floor and the patient is seated comfortably in front. The main tool of diagnosis is to feel the pulse. It is said that a Vaid or Hakim can diagnose practically everything from feeling the pulse. At the most a patient may be asked to stick out the tongue. Rarely, and only for specific problems such as a twisted ankle, would a doctor touch a patient. A patient is never asked to lie down or take up an undignified position. Internal examination is absolutely unheard of. A patient may be touched or massaged for the treatment of some ailment for which it is necessary, but intensive probing is never a part of disease diagnosis. In fact the less the doctor meddles with the patient the better he is supposed to be. The best doctors are those who can tell what is wrong by just 'looking at' the patient. In the western system of medicine the patient is made to feel as if he or she is totally at the mercy of the doctor. Most of the diagnosis involves a physical examination in which the patient is made to lie down and the doctor stands towering above. There is little eye to eye contact with the patient as happens when both are sitting, facing each other. Moreover in hospitals it often happens that not one but a team of doctors examine a patient, often discussing the patient among themselves as if s/he does not exist or is not a person. Many patients who have visited hospitals said that they were made to feel like the Indian equivalent of a guinea pigs, a 'bakra' or sacrificial goat.

Western medical system and the Indian people

The alien or the Angrezi, is defined not so much in terms of its objective qualities like 'locality' or mode of transmission, as oral or written, but in more esoteric and abstract terms relating to power and cultural compatibility. Thus what may be viewed as indigenous or 'mine' is what is more easily comprehended within one's own value system. It is understandable that most persons given a choice would not opt for western medicine given the conflict with the Indian worldview. But the situation is more complex. The western medical system was introduced into India by the British as rulers. For the average Indian conditioned by the values of a feudal society, the ruler was next only to God. The ruler was also viewed as a parent figure who ideally held the welfare of his subjects as a divine mission. This is manifest in the often-used term 'Mai Baap' (mother and father) to address a ruler or even a person in power.

In the state of Bikaner, in the feudal region of Rajasthan I was told that all the men had to shave their heads as soon as a ruler died, indicating that they were the spiritual sons of the ruler. 'Whenever a ruler is terminally ill all the barbers start sharpening their razors.' Anything the British introduced, was viewed by most common people in India as done with the aim advancing the common good. People accepted most of the things the white rulers introduced without much debate. But in a vast country like India it was impossible to reach all the places with modern medical services. Initially, western medical facilities were available only to a select Indian elite. And they had to be administered by a white doctor. Western medical facilities were regarded by most people, for decades as a luxury of the rich and as a symbol of the powerful elite. This corroborates with the use of the term 'Angrezi' for it. The rich and powerful used it as a mark their high status. It remained so even several decades after the Independence of the country. Most poor people thus had not much choice as to opt for alternatives in medical care. Most have to make do even now with what ever is available. Lack of alternatives may be one reason for the continuity of traditional health care systems especially in the rural and backward pockets. But for the elite it could, as we see presently be turned into a game of power and identity.

The British introduced their educational system to India too, creating a western educated elite who led the country in every field in the twentieth century. Most of India's nationalist leaders were from these strata. Thus, although they had developed a concept of nationalism and patriotism, which in itself was a western product, they remained western in their outlook. The first Indians to get trained in western medicine well as their patients came from these strata. In the days of the British rule, the ultimate in medical care was to bring in a European doctor or the Indian equivalent, namely a doctor trained in the West. Many considered it essential to go to England for a medical degree, even after the first medical colleges were opened in big cities like Calcutta and Lahore. In the Fifties educated Indians would scoff at the idea of going to a, what they would call a 'non-medical doctor' meaning thereby a person not trained western medicine. Even today one will find looking at the degrees of doctors in any hospital that there are many foreign ones like FRCS, MRCP etc. among the senior doctors in the fifty plus age group.

'Angrezi Dawai' both symbolized superior power and alien knowledge. While the knowledge, and its practice, were held in a mixture of awe and suspicion, its use was related to an acceptance of the superiority of the 'Angrez' or the British by the fact that they were rulers. In the Pre-Independence era, Mahatma Gandhi thought it appropriate to defy the British by rejecting their systems of cure. Gandhi's practice of folk medicine is well known and so is his rejection of the use of the hypodermic needle as 'violent'.

The political elite and indigenous medicine

One might suppose that by the twenty first century, given the large number of medical colleges in India and the number of medical practitioners they train, every person practically in the sub-continent would have switched over to consulting western medicine. Instead one finds a proliferation of what among the elite are known as 'Alternate Systems of Medicine' and for the poor and illiterate, represent a renewed faith in their traditional systems of medicine.

The transition is a response not so much to indigenous as global political and economic changes. With the demise of colonial powers and the replacement of Europe and especially Britain from its throne of world ruler, many changes have occurred in people's perspectives. It is as if suddenly many people realized that the west does not have a monopoly on wisdom and of power linked to that wisdom. Many people believed in evolutionary philosophy that placed Western civilization at the apex of cultural progress. The highest social attainment for the elite of India in the Nineteenth and early Twentieth century was to replicate the English gentleman's style of life. Children from affluent families were either sent to England or were educated in anglicized schools in India. The acquisition of the correct British accent and table manners were highly valued. The new Indian elite have turned away from copying the so called colonial mentality or slavishly pursuing all things western. The contemporary Indian elite feels proud to extol the virtues of its culture's age-old wisdom and to resurrect its ancient systems of knowledge which the colonizing powers had categorized as superstition and hocus-pocus. Sillitoe (2000: 9) describes the same process for other Asian countries as well.

This ethnic revival movement is not of course limited to medical systems. It features many dimensions including fashions, music, dance herbal beauty aids. In the medical system it manifests a most powerful dimension which deals with deeper convictions and not merely fads and fashions. The movement is part of a new discourse, which has emerged in the eighties and through the nineties, contrasting the scientific with the non-scientific.

During the colonial period the western educational system indoctrinated people, that 'science' was synonymous with west. As was the modern and progressive. The post-modernists have thoroughly critiqued this view. We need to understand the historical and social circumstances that brought about a change in the attitude of Indians who have not been reading the post-modern critiques. A large number of middle and working class Indians have turned away from Western medical systems to rely more of the traditional practices. In other words it would mean that they were no longer equating 'scientific' with 'western' only but with Indian as well.

Disillusionment with the western

Westernization continued to be viewed as synonymous with development and progress until about the mid-seventies. The change was gradual. The first crucial issue was the forced sterilization conducted under the regime of Indira Gandhi, as a part of the Internal Emergency declared by her during 1975. For the first time people started regarding the hitherto 'God' like doctors as public enemies. There were numerous often mythical tales about the horrors of forced or even voluntary sterilization on men and women. Most persons I met during this period (I was conducting fieldwork among the Dhobis of Old Delhi) complained of the loss of manhood and energy and among women, excessive bleeding, pain and infections. Both men and women subjected to the operation complain that they have never been quite the same since. People became fearful of visiting hospitals and being sterilized unsuspectingly. Stories abounded of young, unmarried boys going to hospital with only a sore throat and being sterilized. The modern medical system leaves the patient with little volition or control. Doctors misused the control that they had over you in the hospital.

The second prominent issue that contributed to disillusionment with western science concerned many of the big dam projects criticized for their environmental damage. The first of these to come to public attention was the Silent Valley project, which was halted by Indira Gandhi herself. Mass movements against other such projects such as the Narbada valley and the Tehri dams began soon after. There were grassroots environmental movements (Vervoom 1998: 174) that complemented well with the changing political atmosphere.

The change in ways of thinking was slow but steady. Initially, it was not so much a change from western to indigenous systems of medicine but as the latter acquiring a status akin to the former. In other words the educated elite was now prepared to grant science like status to indigenous systems of knowledge. This certainly does not amount to a rejection of the western system, rather an elevation of the indigenous systems to the same level of acceptability. Although it may seem that indigenous systems are used only for ailments not really considered 'serious' there is a movement towards resorting to these in more drastic illnesses too, such as cancer. For some diseases that are incurable by modern medicine there are many stories of miracle cures through indigenous systems. The press has played a significant role, rather than dismissing practitioners as quacks, as previously. It seeks to bring to light these 'undiscovered' sources of knowledge. For the rural and not literate this means that there is a greater acceptability of their systems and so they can reaffirm their faith in their own knowledge. It also means a consolidation of traditional identities, which is being increasingly felt both the socially and politically.

Although responsible for forced sterilizations, it is to the credit of Indira Gandhi that she spearheaded the ethnic revival movement among the Indian intelligentsia in the post Independence era. In the days of the freedom struggle of the Indian people it was Mahatma Gandhi who had realized the economic potential of the small scale handloom industries for the rural poor. Indira Gandhi was responsible for reviving the attraction of Indian handlooms among the elite and creating a sense of pride in Indians about Indian culture. However it was not her intention to introduce Indian medicine to the people. Science was left untouched as a coveted western thing, unlike culture. Western science together with Indian culture was the right mix for the elite. But the masses were already fearful of western science. The negative aspects of western science had become all too apparent with sterilization and the loss of their lands to dams.

The time was ripe for the new political elite to do a turnover and opt for a different approach to the 'scientific'. This was further made possible by a change in the constitution of the political elite itself.

The new political elite

The politicians who ruled the country just after Independence were highly westernized having received their formal education in largely western institutions. Over the years however a new generation of homespun leaders have gradually replaced them. A significant change was the rise of so-called grass roots politicians, people from the masses with sometimes very little formal education. There was also a great change in the caste factor. The high caste, Brahmins were slowly replaced by a first or second-generation political elite belonging to the lower castes. Many of them came from rural backgrounds, from regions where little westernization had penetrated and who were schooled in the local languages and parochial traditions.

They wrought a significant change in ideology. From the worship of western systems ways and knowledge, there was a revival of the Indian ways and systems of thought. The present Bharatiya Janata Party (B.J.P.) government although branded as a Hindutva[5] party has definite ideas of patriotism and nationalism. This political change was matched by a social change where a new generation of Indians grew up as 'the children of free India'. There was a distinct transition to 'Indianness'. The westernized ways of the pukkah sahibs were also lost no doubt due to global transformations with the popularity of Indian cuisine and Punjabi pop singers in the west.

People questioned the privileging of western ways of life and knowledge. Money was invested in resurrecting and granting respectable status to indigenous forms of medicine such as Ayurveda. Colleges of Ayurvedic medicine, such as the one at Hrishikesh, were opened and award recognized degrees in alternative forms of medicine. In a recent move the Delhi Government formed by the Bharatiya Janata Party has introduced departments of homeopathic and alternative systems of medicine into government run hospitals, though not without some protests from the medical profession. Well-respected doctors of alternative medicine are being given a chance to practice at large hospitals, both private and government run. A front page headline in the *Times of India* (May 2, 2000) reads,

> PM embraces ayurveda
> NEW Delhi: Prime Minister Vajpayee seems to be the latest convert to ayurveda. Recovering from a severe throat infection that made him alter his schedule last week, Vajpayee told a gathering of Ayurvedic experts that he tried allopathic initially which did not help him.
> Now I am taking an ayurvedic medicine which seems to be curing my problem and hopefully by tomorrow I will be fine, he said.
> The government proposes to set up an ayurvedic hospital to promote the indigenous system of medicine in the country. Vajpayee said after giving away the Ramnayarayan Vaidya awards for ayurveda here on Monday.
> It is good ayurveda is becoming popular not only within the country but outside too Vajpayee said.
> He said the budgetary allocation for ayurveda had been doubled this time to Rs. 100 crore[6] and a medicinal plant board be set up.

This clipping shows clearly the use of the term indigenous knowledge and its political implications. It shows the symbolic assertion of an Indian identity by a prominent person. It also reveals an aspect of the game of power. The Prime Minister, who belongs to the Bharatiya Janata Party, a party known for its nationalistic and Hindu sentiments, is making a special effort to promote Ayurveda, not the Hakimi or the Unani system of medicine, which is identified with the Muslims. In identifying indigenous knowledge with the Vedic, Hindu traditions, he also emphasized in a covert fashion, the Hindu Indian identity. While the common people include Hakimi and Homeopathic, as well as 'Jhar Phoonk' within their definition of Desi Dawai, the government is at pains to make indigenous knowledge synonymous with the classical Hindu Indian tradition.

Thus the idea of indigenous knowledge as 'our knowledge' is being used in India as a political weapon, as a means to enhance a sense of 'we' ness, in opposition to the now discredited 'Angrez'. In a subtle way the politicians are also trying to control the definition of what 'we' means. In the B.J.P. context it means Hindu and high caste. This emerging identity is an aspect of resurgent patriotism as symbolized by the midnight celebrations at India Gate, in New Delhi; to celebrate Fifty Years of India's freedom, thronged by a million strong crowds. In a less obvious fashion the concept of indigenous is also being used to define what is 'ours', that is building up on a specific communal and class identity.

At the local political level the acceptance of local knowledge systems gives a sense of power which can be used in political rhetoric. Small political movements sometimes copy what is happening at the larger national level. Over the past decade for example the people of the Himalayan region of Uttar Pradesh have been demanding a separate identity as a hill state Uttarakhand.[7] This reflects people's resentment at the imposition of plains people culture on the hills. Recently, following the revival of the traditional medical systems, little tents have proliferated all around Delhi, near the busiest traffic intersections, all peddling Himalayan herbal cures by specialist Vaids from the mountains. This knowledge of medicinal herbs, accessible only to the people of the mountains, is one way for them to exert intellectual hegemony over the people of the plains. It is a political statement that they have a valuable system of knowledge.

Indigenous knowledge is understood here not as a pragmatic path to aid 'development', or as a rational system of knowledge, but as a symbol of power, assertiveness and identity. In a way the manner in which indigenous knowledge is being used by this de-colonizing society today is no different from the way its opposite 'scientific knowledge' was privileged by the colonial powers. Science was used as a symbol of their superior 'power' presented as superior 'knowledge' and more dangerously as 'universal' knowledge in the sense of undisputed truth. Even today anthropologists pay lip service to this colonial hegemony by dubbing indigenous as 'local'; thus concealing the conditions under which once local western science became global; featuring violence and dominance as illustrated by Taussig (1987).

If people are using indigenous knowledge today it is because the 'colonized' have realized at the level of both political elite and at grass roots that they can achieve an identity only through asserting their freedom and becoming 'de-colonized'.

NGOs and grass-roots movements

The eighties saw a resurgence of NGO movements, many at the grass roots. They were committed to a different development approach than of the Government, the latter based on colonial western science and modernization. The NGOs sought to reverse this and promote a more indigenous version of development. One such NGO in Rajasthan headed by a woman serve to illustrate their ideological and practical aspects of development and the impact they are having on people. All those involved with this NGO were local and although they had not received what is known as a western education, they had all studied in the vernacular but were well educated. The main aim of their work was to instil sense of self-worth and pride in the local people. They actively campaigned against a dam that was to be constructed in their area, which has yet to be built, and a hotel on a small island hitherto inhabited by only a handful of tribals. Both these projects were however completed eventually.

One of their strategies to end unwanted westernization was to affirm the efficacy of traditional medicine and warn of the dangers of western medicines and systems of cure. The works of this NGO is two folds. Firstly, it has organized a massive propaganda campaign against westernization, linking it to the exploitation and patenting of the indigenous resources such as medicinal plants and herbs by multinationals and the alienation of the local herbalists physically from their plants, by the creation of reserved forests. Secondly it seeks to reinstate the traditional practitioner by elevating him to the status of a scientific practitioner of medicine. One of the NGO's programmes formally recognises traditional medical practitioners, designated as a 'Guni', which means knowledgeable or talented man. They searched for such 'Gunis' from among local practitioners who they considered as authentic. On one of my field visits I observed them making a local Ayurvedic preparation called Chyvanprash, with some local Gunis and village women, which they then packed in cans and sold to the local people at a reasonable price.

The NGO is fighting at the ideological level, questioning the privileging of western science over indigenous systems of knowledge, and at the practical level, recreating a new status, the Guni, who should be respected, like the practitioners of modern science. They were often at odds with other NGOs run by western funds and agencies that privileged western science over the indigenous and the Government programs too. These latter aimed to get western medical practitioners and medicines to the people but failed because of the apathy of the local doctors trained in formal medical schools. This strengthened the platform of those advocating indigenous systems of knowledge who asked: 'what do we get except out dated medicines and untrained nurses? It is better to rely on our own resources and develop them'. It is appropriate to consider the factor of poverty for many have no choice but to make do with what ever is available. But the important trend is a turn around by those who have a choice, that is those who can afford any kind of treatment but still prefer to opt for indigenous ones. At one meeting to honour local Gunis, they had several western trained medical practitioners (i.e. having MD degrees) present to highlight the achievements of the indigenous system of medicine. This signifies the change occurring in the attitudes of the elite, medical practitioners and the government to support the shift towards alternative or indigenous systems of medicine. Thus the political use of identity is being made by those who have power and not by those who are too poor to make any change in the system. Even the NGO are only making use of these people to achieve their ideological ends.

The significant contribution of the grassroots NGOs is their awareness of the likelihood of 'decivilisaton, depoliticisation and dispossession' (Fairhead 2000: 100) if the government is allowed to take over completely the process of development. It is this that they are fighting against and the concept of indigenous knowledge has become a powerful tool in this fight.

At this point it is relevant to consider the relevance of attempts by NGOs and others to give respectability to indigenous knowledge by formalizing it in the fashion of western knowledge. It is at this point that the failures arise for as realized by most such people it is not possible to formalize and standardize indigenous knowledge in the manner of western knowledge (Ingold: 1997; Sillitoe 2000: 11) nor is it necessary (Aggarwal 1995: 430). The manner in which indigenous knowledge is acquired, by application rather than by formal communication makes it nearly impossible to codify. Thus the actual use of indigenous knowledge remains limited but what becomes important is its use as a political statement. For to really adapt indigenous knowledge what one would have to acquire is the 'disposition' that goes with it (Ingold 1992, 1997: 75).

The market and commercial interests

There has been a commercial aspect to this change The market plays a role, as pointed out by Sillitoe (1998) by increasing alternatives and the levels of consumption. The indigenous knowledge movement has seen many Indian companies jump onto the bandwagon, not for patriotic but for purely commercial reasons.

Indian drug companies have profited from the changing climate such as the Dabur and Himalaya Drug companies that are marketing patented Ayurvedic drugs in every major drug store around the country. To add to their respectability many allopathic doctors are prescribing some of these drugs in a big way especially for those ailments for which they have no other drug available. Prominent among them are drugs for liver ailments, rheumatism and tonics for old people. Medicines for infections and menstrual cramps are also highly popular as are those cosmetic problems of skin and hair. Interestingly these drugs gain greater respectability if they are accepted by western medicine. The advertisement for a recent series of Ayurvedic medicines for example, shows an elderly traditional lady introducing such medicines to a group of 'modern' that is westernized teenagers in a fast food outlet winning their rapt attention by speaking English and using technical, scientific words. The message is clear: the indigenous is as modern and scientific as the western, which increasingly is being seen, as American rather than English.

Conclusion

Post-modernism involving itself a critique of western rationality has opened up a debate on the hitherto unchallenged status of western science. People in India engaging in their own debate are also questioning the hitherto 'superior' western knowledge, especially in medicine, turning to more indigenous systems of knowledge regarding disease and cure. This shift is not due to any one single factor but the complex interplay of several social and political ones including an identification of western systems of knowledge with colonization. Nationalism promoted by a homespun political elite is prompting the current generation of Indians to exert their identity through a reaffirmed faith in their own knowledge systems (Vervoorn 1998: 172).

Jan Mohammed (1986: 80) distinguishes 'dominant' and the 'hegemonic' forms of colonialism. By 'dominant' he means 'the period from the earliest European conquest to the moment at which a colony is granted 'independence', European colonizers exercise direct and continuous bureaucratic control and military coercion of the natives: during this phase the 'consent of the natives is primarily passive and indirect'. The second phase of 'hegemonic' is when 'the natives accept a version of the colonizer's entire system of values, attitudes, morality, institutions, and, more important, mode of production. This stage of imperialism does rely on the active and direct 'consent of the dominated' (Jan Mohammed, 1986: 81). The acceptance of the Western science as 'truth' by most Indians resulted from hegemonic colonialism. However the basis of colonialism, mainly education, was also its undoing. Today it is the upper classes, the elite, who are keen to reject the western system for indigenous knowledge. The acceptance of indigenous knowledge symbolizes the rejection of colonialism as well as a process of enlightenment to combat 'hegemonic' colonialism. The British made their system of education THE system of education for elite Indians to the exclusion of all traditional forms of schooling and education. The latter were left for those who could not afford the more expensive western education. Ironically it was the elite, tutored in the Western traditions of individualism and personal freedom, which was overtly instrumental in rejecting 'dominant' colonialism. The uneducated rural masses did not distinguish between one class of power holders and another. For them the feudal lords, the 'Mai Baap', were replaced by the British, who were replaced again by the rulers of democratic India. It was one paternalistic figure replaced by another. Few of them were exposed to any other kind of knowledge than their own traditions, at any time during their history. India being a vast country, only parts of it had been exposed to the British education. Even today, many interior pockets have no proper school. Even the missionaries had not reached to every corner concentrating more on the non-Hindu tribal pockets like the North-eastern region. The power games excluded a majority of Indians. The colonial power holders had discredited indigenous knowledge and deemed it inferior to the western systems of knowledge.

The post-colonial power holders are now trying to resurrect indigenous knowledge in a bid to over throw remnants of earlier hegemonic colonization and replace it with their own symbols of power (Cheater 1999: 4).

The political classes are using indigenous knowledge in the same manner as Hindutva, to create an Indian nation finally free of all traces of 'hegemonic colonialism', fifty years after its release from 'dominant colonialism'. The uncomfortable fit between western medicine and the Indian ethos is just one reason for its rejection today. Those people who swore by western medicine earlier are highlighting this reason anew. It is not only political interest. Many favour indigenous forms of medicine on moral and value grounds as more appropriate. For the poor of India who have neither access to, nor the means to make use of expensive western medicine, it has remained the only viable alternative. The approval given to indigenous knowledge by the powerful enhances it value for every one. The democracy gets strengthened as the poor and earlier marginalized feel drawn into the mainstream. This definition of indigenous knowledge as 'mine', as part of my identity and power is quite different to technical use in development discourse. In India indigenous knowledge is understood not only as knowledge but also as symbol of nationhood. However the manipulations of identity are only tools for the powerful. The poor remain where they were: outside of all discourses on power.

Notes

1 Sillitoe in a personal communication assures that he was speaking as a natural scientist (agricultural scientist) and that there are many such Asian scientists (the 'us' in this sentence) . And to be sure in the UK there are IK too e.g. the farmers of the UK as against the scientists. However I also meant it in this way that the divide exists between Western science and the non scientifically trained person, not that between Asians and

Westerners. The Asian who qualifies to be called a scientist is a person trained in western science. A person who does not have such training, whether Western or Asian is not called a scientist, in spite of whatever knowledge possessed by him.

2   There is possibility of a disagreement here and some like Sillitoe may not be of the view that the indigenous always refers to the Other. However as Beteille and many other would agree, it does in most cases does assume a non-European culture.

3   I have used Hindi terms as it is the national language and also most widely understood in India. Otherwise India has fourteen official languages and numerous dialects.

4   The term allopathic is universally used in India to refer to the scientific or western system of medicine the kind practised by the holder of a M.B.B.S degree. I do not know the historical reason for this But all Indians refer to it as such and hence I have used this term in this chapter.

5   The term Hindutva refers to a Hindu revivalism currently under way in India.

6   One crore rupees is ten million rupees, approximately £150,000.

7   This state is now in existence as the state of Uttaranchal.

References

Agrawal, A. 1995. Dismantling the divide between Indigenous and scientific knowledge. *Development and Change* 26: 413–439.

Alter, J.S. 1999. Heaps of health, metaphysical fitness: Ayurveda and the ontology of good health in medical anthropology. *Current Anthropology* 40. Supplement: 43–66.

Arce, A. & N. Long. 2000. *Anthropology, development and modernities: Explaining discourses, counter-tendencies and violence.* London: Routledge.

Beteille, A. 1998. The idea of indigenous people. Current Anthropology 39(3): 187–191.

Bourdieu. Pierre. 1972. Outline of a Theory of Practice. Cambridge: Cambridge University Press.

Channa, S.M. 1993. Value oriented development: A case study related to development of indigenous health practices. *Social Action* 43: 495–500.

Chatterjee, Partha. 1989. Colonialism, nationalism and colonized women: The context in India. *American Ethnologist.* 16 (4): 622–6.

Cheater, Angela P. 1995. Globalising and the new technologies of knowing: Anthropological calculus or chaos. In *Shifting contexts: Transformation of anthropological knowledge.* (ed.) M. Strathern.Routledge: London & New York.

Cheater, Angela P. 1999. Power in the Post-modern era. In *The anthropology of power.* (ed.) A. Cheater A.S.A.Monograph. London: Routledge.

Cohen, A.P. 1993. Segmentary knowledge: A Whalsay sketch. In *An anthropological critique of development: The growth of ignorance.* (ed.) Mark Hobart. London: Routledge.

Fairhead, J. 2000. Development discourse and its subversion: Decivilisation, depoliticisation and dispossession in west Africa. In *Anthropology, development and modernities.* (ed.) A. Arce & N. Long. London: Routledge.

Ferradas, C. 1998. Comments on Paul Sillitoe's paper. *Current Anthropology.* 39 (3): 239–240.

Foucault, M. 1972. *The archaeology of knowledge.* New York: Pantheon Books.

Foucault, M. 1980. *Power/knowledge. Selected interviews and other writings, 1972–1977.* (ed. & trans.) C. Gordon. Brighton: Harvester Press.

Gledhill, J. 1994. Power and its disguises: Anthropological perspectives on politics. London: Pluto Press.

Hobart, M. 1993. Introduction: The growth of ignorance. In *An anthropological critique of development: The growth of ignorance.* (ed.) Mark Hobart. London: Routledge.

Ingold, T. 1992. Culture and Perception of the environment In *Bush base: Forest farm: culture, environment and development*. (eds.) E. Croll & D.Parkin. London: Routledge.

Ingold, T. 1997. The optimal forager and economic man. In *Nature and society: Anthropological perspectives*. (eds.) P. Descola & G. Palsson. London: Routledge.

Jan Mohamed, A.R. 1986. The economy of Manchean allegory: The function of racial difference in colonialist literature. In *Race, writing and difference*. (ed.) H. Gates Jr. Chicago: Chicago University Press.

Jordanova, L.J. 1980. Natural facts: A historical perspective on science and sexuality. In *Nature, culture and gender*. (eds.) C.P. MacCormack & M. Strathern. Cambridge: Cambridge University Press.

Kakar, S. 1982. *Shamans, mystics and doctors*. Delhi: Oxford University Press.

Kakar, S. 1986. Psychotherapy and culture: Healing in the Indian tradition. In *The cultural transition: Human experience and social transformation in the Third World and Japan*. (eds.) M. I. White & S. Pollack. Boston: Routledge & Kegan Paul.

Lyons, M.L. & J.M. Barbalet. 1994. Society's body: Emotion and the somatization of social theory. In *Embodiment and experience: The existential ground of culture and self*. (ed.) T. J. Csordas. Cambridge: Cambridge University Press.

Moore, H. L. 1994. A Passion for difference: Essays in anthropology and gender. Cambridge: Polity Press.

Pottier, J. 1993. *Introduction*. In *Practicing development: Social science perspectives*. (ed.) J. Pottier. London: Routledge.

Pottier, J. n.d. Modern information warfare versus empirical knowledge: Framing the crisis in Eastern Zaire. 1996 Draft paper presented at ASA 2000 Conference, SOAS, April 2000.

Sillitoe, P. 1998. The development of indigenous knowledge A new applied anthropology, *Current Anthropology*. 39(3): 223–252.

Sillitoe, P. 2000. The state of indigenous knowledge in Bangladesh. In Indigenous knowledge development in Bangladesh: Present and future. (ed.) Paul Sillitoe. Dhaka: The University Press.

Strang, V. n.d. Close encounters of the third world kind: Indigenous knowledge and relationship to land. Draft paper presented at ASA 2000 Conference, SOAS, April 2000.

Taussig, M. 1987. *Shamanism, colonialism and the wild man*. Chicago: University of Chicago Press.

Vitebsky, P. 1993. Is death the same everywhere? Contexts of knowing and doubting. In *An anthropological critique of development: The growth of ignorance*. (ed.) M. Hobart. London: Routledge.

Vervoorn, A. 1998. *Re orient: Change in Asian societies*. Oxford: Oxford University Press.

Chapter 10

In Dialogue with Indigenous Knowledge: Sharing Research to Promote Empowerment of Rural Communities in India

R. Baumgartner, G.K. Karanth, G.S. Aurora and V. Ramaswamy

'You researchers, you kindly care for your research and let watershed development be my business!' This was the harsh reaction of a local contractor demanding 'role clarity' at an assembly of villagers who had come to listen to a research team present their findings from a study on changing rural livelihoods and natural resource management in their hamlets. Yet, it was not the research feedback that provoked his reaction, but the invitation: 'Do you want to know how watershed development occurred in my village?' extended to the research team by a middle-aged woman, courageously drawing attention to her experiences of a government initiated watershed programme in her village. In her words, she only realized that a watershed project had started when – without prior consent and without recognizing a specific need – civil contractors were about to erect check-dams on her private land. A village watershed committee had approved and signed a watershed development plan, but the committee had been formed unknown to the villagers in a clandestine agreement between the influential local contractors and representatives of the government department concerned, thereby making mockery of regulations to ensure people's participation.

In complicity with government administration, powerful members of the local elite have obviously learned how to divert the growing stream of external resources earmarked for local rural development to themselves. Indeed, the watershed development programmes of Andhra Pradesh State had been under scrutiny for quite some time for systematic misappropriation of project funds (Roopi 1996). The intervention described here not only revealed the local balance of power, but to our own discomfort, also raised the question of whether our feedback session had encouraged an overexposure of a possibly vulnerable woman, who bravely voiced her concerns.

The above example points to the various risks involved when providing research feedback to rural communities. This chapter explores the scope and limitations of promoting mutual learning through research feedback. It focuses on two South Indian villages where we conducted development-oriented research for over two years. It aims to share the practical experience and lessons learnt from providing research feedback in the context of participatory field studies.

Vision and reality of participatory research

The basic assumptions for fostering mutual learning in field research are straightforward. Mutual learning requires participation, turning away from data mining research to partnerships between research teams and rural communities. The research agenda should reflect the balanced interests of both partners. Promoting shared ownership of research results is the most challenging aspect of such an approach, which also faces awkward questions about research – for example, responding to farmers who increasingly challenge researchers with legitimate queries such as: 'How does your research benefit our community or our household?', and also convincing development agencies that question the usefulness and relevance of such research to rural development. Through such enquiry we are made increasingly aware of the gap between the vision and the reality of participatory research.

Although a more participatory approach to rural development research has been called for in the development literature for some time (e.g. Cornwall & Jewkes 1995; Chambers 1979), progress from rhetoric to meaningful participation has been slow and difficult, even in programmes that place it at the top of the agenda (see Hinchcliffe et al. 1999; Rhoades 1998). It is not yet established practice for researchers to feel accountable to the 'subjects' of their field research or, as John Clammer (2002) puts it: 'Contributing to the development discourse requires more that disseminating indigenous knowledge in a complex professionalized form and scientific language.'

Moreover, we have to revise some assumptions such as the notion that all communities are homogeneous and hence participation can follow a uniform approach. Rural realities are to the contrary. Research and development must cope with complex livelihood patterns within the same community, and risk getting caught in a web of factional interest and overlapping power relationships (see Jeffery & Sundar 1999: 37-40). We should, however, not allow such problems to deter us from attempting to engage in actor-oriented research, especially for supporting participatory development collaborations.

Research context

The Indo–Swiss collaborative research project on 'Rural Livelihood Systems and Sustainable Natural Resource Management' (henceforth, RLS Research) is an 'actor oriented field study'.[1] It is based on the assumption that farmers and their communities have developed culture and location specific patterns of land use. However, from their perception land use represent just one, albeit important, dimension of a much wider concern on their part for sustainable livelihoods, demanding a constant adaptation of survival strategies.

This project assumes that the household strategies of farmers are primarily aimed at sustainable livelihoods and not merely at sustainable use of a single natural resource such as land, water or forest. The worldviews that inform this goal vary greatly. For example, a development project to improve land use among Bhils tribes people in Gujarat (see Koppers 1948) will have to contend with a livelihood strategy featuring seasonal migration to survive on ancestral land, quite different from an increasingly market-oriented Reddy landholding caste community in Andhra Pradesh (e.g., Aurora 1999). The RLS project was designed to explore changing livelihood strategies in rural India and study the impacts on the management of natural resources.

From 'knowledge for understanding' to 'knowledge for empowerment?'

The RLS research aims to go beyond the conventional goal of 'knowledge for understanding' to generating 'knowledge for action' (Scott & Shore 1979). The latter is necessary for supporting more effective programmes of sustainable rural livelihood development that avoids degrading the natural resource base. Accordingly, research feedback serves as a platform for discussing controversial issues related to processes of resource degradation identified by the project, and which may be relevant for any later involvement of the people in development. The feedback also clarifies questions such as the following: has the research really captured the farmers' own perceptions and understandings of resource management, and has it produced new insights within the community regarding future development of their natural resource base? Does involvement with research contribute to the community's competence and bargaining power in future interactions with outside development agencies? In short, does the participatory interaction between researchers and villagers promote creative change, advancing local community empowerment (Nelson & Wright 1995)?

In the absence of any established concepts, the feedback attempted by the RLS research offered many conceptual challenges. Before relating our approach and experience, we first address some critical aspects of participatory research and development in the Indian context and take a clear stand on some controversial issues.

Towards a livelihood perspective in participatory research

There is a consensus among development agencies that active participation of rural communities is a prerequisite for any development programme aiming to advance sustainable village management of natural resources (Kuhn 1998). Rural households will participate in sustainable resource management projects only if it connects meaningfully with their livelihood concerns (Högger 2000). Such concerns are not limited to the economic, but have social, spiritual, and emotional dimensions; they are holistic in nature. How does one acquire a holistic understanding of livelihood in field research? The RLS project found an answer in the interface of two powerful images used for a holistic perception: the nine-square mandala as a cross-culturally accepted symbol for wholeness and a centred universe, and the rural house as a metaphor for livelihood (see Högger 2000). Figure 10.1 shows the adaptations for the RLS research.

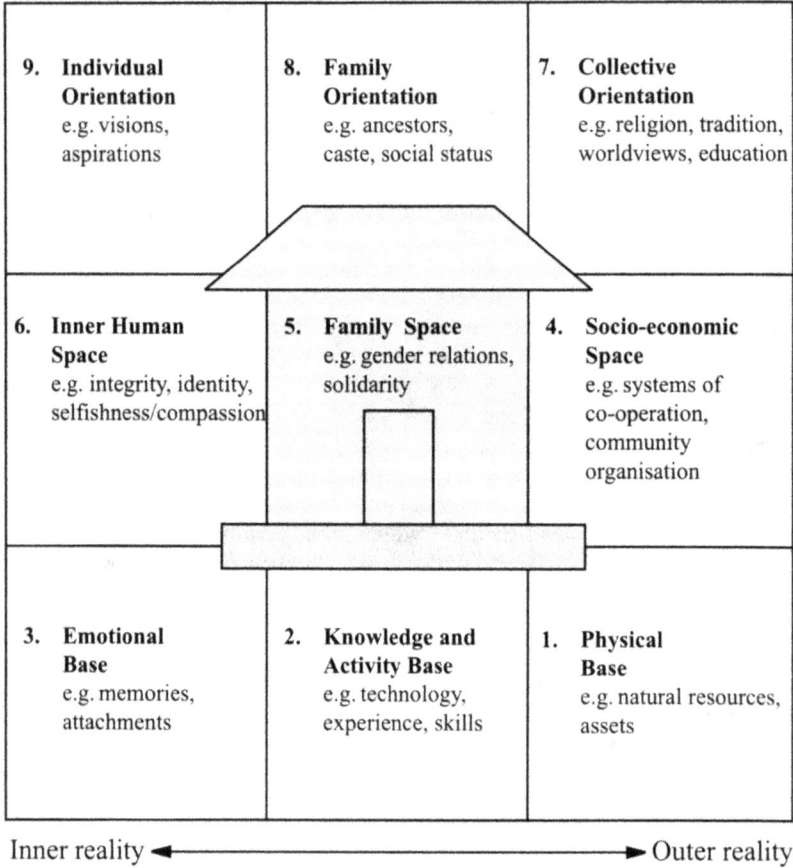

Figure 10.1 shows a nine-square grid arranged as a house with roof, walls, and foundation:

9. Individual Orientation e.g. visions, aspirations	8. Family Orientation e.g. ancestors, caste, social status	7. Collective Orientation e.g. religion, tradition, worldviews, education
6. Inner Human Space e.g. integrity, identity, selfishness/compassion	5. Family Space e.g. gender relations, solidarity	4. Socio-economic Space e.g. systems of co-operation, community organisation
3. Emotional Base e.g. memories, attachments	2. Knowledge and Activity Base e.g. technology, experience, skills	1. Physical Base e.g. natural resources, assets

Inner reality ◄───────────────────────────► Outer reality

(RLS Högger 2000)

Figure 10.1 Livelihood seen through the nine-square Mandala

In Figure 10.1, the nine square mandala differentiates between an inner and outer reality of livelihood. The metaphor of the house suggests a three-tiered perception of livelihood where the foundations represent the material and non-material resource base while the walls create room for three different notions of 'space'. The roof finally points to the three-fold orientation of a livelihood system. In this application, the mandala also conveys the notion of interconnected knowledge domains in a livelihood system. Yet – and this is the practical implication for development interactions – it offers us the choice of context related entry points to a livelihood system. In other words, it allows a combination of a fragmentary approach with a holistic perspective! (see Carney 1998).

An irrigation tank in the interface of diverse livelihood perspectives

Rural livelihoods are diverse and shaped by culture and location specific conditions and experience (Jodha 1990). Different persons may perceive a natural resource differently. We shall illustrate this by looking at a typical irrigation tank in a village on the dry plateau of the Deccan in South India. While walking on the dam of such a tank, researchers or development advisors may be intrigued by the extent to which the tank bed has silted up, seeing that the tank can no longer irrigate the original command area. Cow and buffalo may even graze the dry and sandy tank bed. The upper catchment from where the tank draws water during the monsoon is equally ruinous: barren, depleted of trees, the few remaining bushes ravaged by fire, the eroding soil a burnt crust. The massive tank bund winding serpent-like across the watershed, the masonry of the sluice gates and the skilfully designed web of irrigation channels – all inspire respect for past generations that erected the irrigation system and maintained it for more than a hundred years.

Most visitors feel attracted by the potential benefit of a rehabilitated tank system. In their spontaneous development vision they envisage replanting the upper catchment with trees, desilting the tank and clearing the canals. A rehabilitated tank and watershed should lead to sustainable grain production, an increase in fodder yield, firewood and timber. Fishermen might return, as might potters who would find access to clay previously blocked by sandy silt, and washermen too. Animal husbandry would flourish with plenty of water in the tank and grass in the upper catchment. And finally, groundwater levels, rising again under the effect of steady percolation, would replenish the open wells in the neighbouring village dry lands allowing irrigation in dry spells. A guaranteed road to sustainable resource management – the eyes of the visitor!

The elderly former village headman, some farmers, a few boys and the village teacher might represent the 'people' to the outsiders strolling over the tank bund. This small cross-section of villagers may have varying perceptions of the tank. The former headman may see bright prospects for a state-supported programme of tank rehabilitation. It would bring work for one of his sons, a civil works contractor. Mobilizing the village workforce to maintain the tank had once been the pride of his years in office a few decades ago. The derelict tank reminds him of a bygone era when it was the chief economic asset of his landholding caste. However, some of the farmers have invested in sinking borewells, in recent years, with subsidized loans from the state, to assure private water supply to their fields. They are self-sufficient and do not need the tank. In contrast, another farmer with land at the dried-up tail end of the irrigation channel would readily commit himself to tank rehabilitation to regain access to water to irrigate his crops. The young boys may instead be dreaming of becoming drivers of heavy Tata trucks or cinematographers for popular cinema or television serials. The teacher assumes that dwindling interest in the tank is due to the Department of Minor Irrigation[2] being nowadays responsible for maintenance.

Development perspectives informed by different knowledge systems

The foregoing imaginary scenario could apply to many parts of rural South India. The group on the tank bund is caught between at least two different perceptions of 'sustainability'. The visitor's attention focuses mainly on the irrigation tank as a natural resource, part of a larger biophysical system, informed by the 'outsider' rationality of sustainable watershed management. The farmers' perceptions, differing among themselves, reflect more complex sustainability concerns, aware of the livelihood strategies of individual households. Such strategies are shaped by culture, history and experience, and are influenced by the rapidly changing socio-economic environment at both micro and macro levels.

The group on the tank bund represents an interaction between two different, yet interlinked, knowledge systems. On the one hand, there are perceptions of watershed systems emanating from scientific knowledge. On the other, there are notions rooted in an indigenous knowledge system with its tacit understanding enshrined in the indigenous society, its traditions, history and belief systems (Karanth 1995; Vasavi 1999). The latter consists of a stock of knowledge that does not necessarily contradict 'rational and scientific' principles of sustainable resource management, yet circumscribes their application in a given indigenous context.

The tank bund group may also represent two contrasting development rationales. One sees project design as the result of deductive scientific reasoning. The other envisions a project evolving along an 'inductive' path, starting from local people's perceptions of sustainable development (Pretty & Scoones 1995). Successful tank rehabilitation requires critical examination from both perspectives, the reconciliation of the diverse interests of multiple indigenous stakeholders with the requirements of scientifically sound ecosystem development (Dixon 1997).

Participatory watershed development in India, which, in recent years, has tried to go beyond the rhetoric of participation, offers insights into clashes with government departments and their 'top down' tradition. It also reveals the many hurdles faced in attempting to circumvent traditional indigenous power structures in the broader interests of the community. 'Stakeholder complexity and competition' is one of the eight operational 'landmines', as Rhoades (1998) calls them, jeopardizing the implementation of participatory watershed management.

The 73rd Amendment to the Constitution of India focuses on people's participation (Oommen 1995), seeking to empower the gram sabha (literally, the village assembly), comprising all those registered in a village's electoral rolls to make collective decisions in village affairs, specifically pertaining to improved use of natural resources and community social and economic development. By taking part in the process of decision making it is assumed that people are empowered. In reality, however, gram sabhas too often represent assemblies of the ruling faction, taking the same decisions as powerful leaders, elected or otherwise. We are witnessing in India growing disenchantment with participation while at the same time people's need for empowerment becomes increasingly evident.

Accordingly, research into natural resource management in villages faces at least two challenges. First, how can participatory research capture people's perceptions of natural resource management without compromising academically sound investigations? Second, how do we design and conduct participatory field research to enhance people's empowerment and promote rural development? We must address these questions with rural communities increasingly aware that we have, in the past, quarried them for data mainly used for academic advancement.

Agenda setting, ownership and empowerment in field research

Recent development literature is rich on reflections on the question 'whose reality counts?' when discussing deteriorating rural livelihoods and setting development goals (Chambers 1995). Figure 10.2 summarizes the basic dimensions of participation.

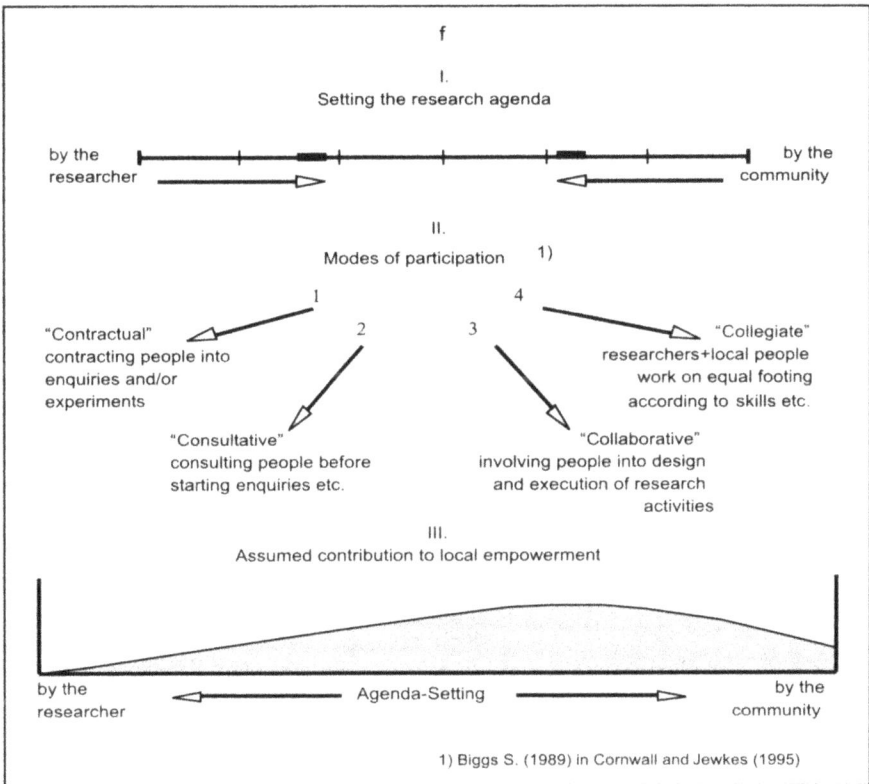

Figure 10.2 Participatory research – from agenda setting to empowerment?

Conventional understanding of participatory research has the researcher as a participant observer spending an extended period in the community studied and experiencing the daily lives of the people to achieve an empathetic understanding. The researcher sets the agenda largely with the community as an object of study.

There is hardly any local empowerment. If, on the other hand, the research agenda is set entirely by the community would this promise optimal empowerment? Probably not. It would mean foregoing the benefits of cross-fertilization with outside ideas. It may also allow the hijacking of the research agenda by powerful stakeholders in the local society. We argue that empowering participatory research must feature joint decision making by insiders and outsiders. Such research labelled 'collaborative' or 'collegiate' (Figure 10.2) is seldom achieved in reality.

Participation alone does not guarantee quality research. Participating dogmatism may harm research. We must relate it to research goals and ensure that it meets basic ethical requirements when communicating with communities (Whyte 1984). While subscribing to participatory research, we still set our project's agenda in RLS largely as researchers. Against our initial intentions, local participation remained overwhelmingly of the contractual kind. An example was the involvement of local youths in tracing selected farmers' life histories and relating these to the eco-histories of natural resources (Ramaswamy et al. 1996). The youths worked alongside the researchers and were paid for their services. Awareness of our unsatisfactory involvement of indigenous stakeholders in the agenda setting and research process informed our strong motivation to ensure dialogue by means of a carefully conceived research feedback to the community.

Preparing a programme for research feedback: basic questions

While planning the feedback agenda, our concern and respect for locally approved dos and don'ts influenced its conception and organization. We asked ourselves if women and men, young or old, of the five hamlets would be willing to attend the time-consuming events termed 'Research Feedback' after having interacted patiently with the research team over a period of two years. The dreaded, worst-case scenario was to end up sitting under a Banyan tree in a hamlet entertaining a few curious children with well-prepared documents and visuals. We wished to comply with villagers' traditions of meeting and communicating with guests, of expressing approval or disapproval.

It was necessary to secure prior commitment to attend meetings from each hamlet and their caste and stakeholder groups. To enhance a positive response, the research team distributed a brief and simple written translation of the major findings to every household, regardless of its literacy. In addition, we arranged the schedule of our meetings to fit with the indigenous rhythms of daily life. We launched the feedback exercise as a significant event, starting auspiciously with religious singing (bhajan), welcome speeches and attractive 'trailers' of the forthcoming presentations and discussions.

There are some practical issues regarding the preparation of a feedback dialogue that we need to address. Our overriding concern was to provide feedback on the management of the village irrigation tank, changing cropping patterns, seasonal and permanent migration, consequences of the dissolving jajmani system,[3] gender roles and education. Discussions within the research team centred on where to restrict participation to sharing only, and where to stimulate dialogue on conflicting perceptions within the community. Such differing perceptions were either a cause or consequence of many interpersonal or group rivalries. We aimed, however, to go beyond just providing new platforms for old battles.

'Reversed Participatory Rural Appraisal' became the guiding principle of the feedback programme. Many research findings required conversion from abstract academic language into a visual form with the use of metaphors and examples. Should we use skilfully drawn pictures displayed on pin-boards or overhead projectors? The success of the event depended on research team members having the necessary public speaking and discussion chairing skills. Test runs were essential and became opportunities for intensive mutual learning within the research team. The rehearsals involved role-plays and friendly clashes over conflicting interpretations, whilst also focussing our minds on problems of interacting with multiple stakeholders in the community.

Indigenous conventions demand that village dignitaries too have a chance to address the gathering. The issues of who to invite, who should speak first and for how long, exercised us considerably. We were concerned not only to keep their oratory to a time-schedule but also to prevent them from either presenting misleading impressions of the feedback event or hijacking it to their own ends. In faction-ridden villages these are sensitive matters.

During Participative Rural Appraisal and research we experienced the very limited opportunities women have to voice their opinions in public, especially in the presence of elder men. Our strategy was to sit a female member of the research team among the local women in their conventional seating arrangements at public meetings.[4] The female researchers would support women at crucial points during the debates without them forcing themselves forward. We involved local teachers and students by awarding prizes for drawing, essay writing and debating competitions, in which they expressed their ideas about village resources, gender relations, and so on.

An effective feedback process opens up new perspectives. But, in our case, it also placed the research team in an awkward position, having limited influence on development programmes in the region. The research sharing also revealed a classical case of 'cognitive dissonance'. Villagers started to voice their expectations of tangible contributions from the research team to community development. It made us aware to what extent participatory research encouraged people to address researchers as if they were members of an NGO or development agency in the planning stage of a project.

Research feedback on water and livelihood

Water is a highly unpredictable resource in semi-arid environments. Surface water supply depends on monsoons. Changing regimes of water management were a central RLS research theme. On the Deccan plateau, elaborate systems of tank irrigation have been the predominant form of surface water management for centuries. They depend on the rainfall from the previous monsoon, unlike irrigation systems in Northern India that rely on the perennial flow of glacial water from the Himalayas. These tank irrigation systems, managed as common property, functioned in a sustainable manner over long periods of time (Wade 1988). A user group, clearly defined by landholdings in the irrigated area, managed the system. Under the leadership of the village headman, it shared responsibilities for maintaining the three central elements of the system: the upper forested catchment, the tank with dam and sluices, and the command area with canal distribution system.

These systems have deteriorated greatly in recent times. The almost derelict tank located in CR-Palli in the Ananthapur District of Andra Pradesh is typical.[5] The tank covers a surface of approximately 40 acres, with a command area of 200 acres and a vast upper catchment extending to at least 5 km^2 of depleted forests. The tank was built long before British rule. Having served the village for more than 200 years, it gradually silted up after independence and for most of the year serves as grazing ground. A large number of wells have been sunk to compensate for its dereliction.

When it came to a survey of causes for the system's neglect – the eco-history of this water source – the villagers confronted us with a range of conflicting explanations. Instead of trying to sort them into right and wrong and convert them into a flow chart linking logical causes with effects, we decided to approach the people with their own and obviously inconsistent assumptions (Figure 10.3).

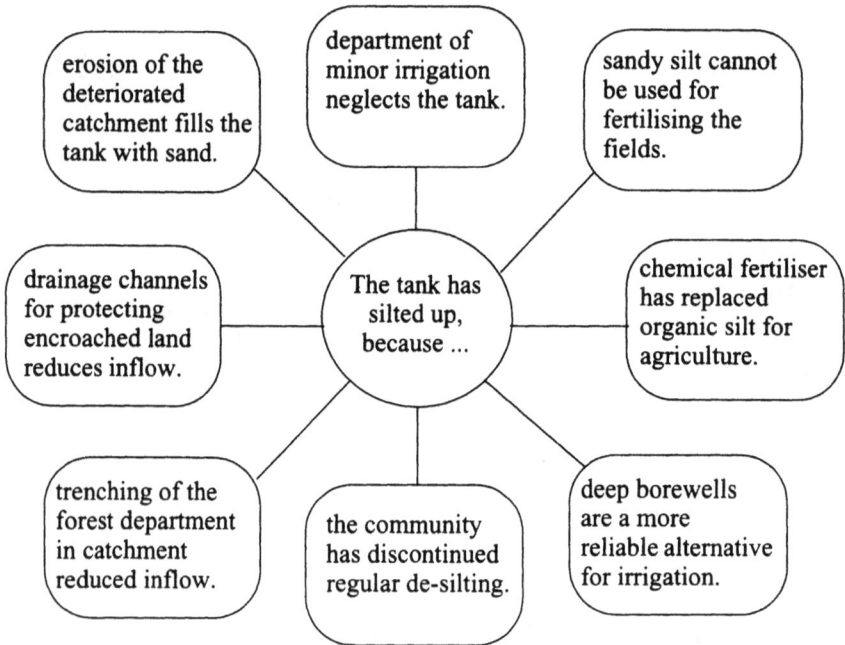

Figure 10.3 Tank management collapsed

How would the five hamlets and the different castes react to these conflicting perceptions, as they emerged from the stakeholder investigation? The RLS project had to meet a methodological challenge with regards to the following:

- to come up with an appropriate arena for community debate;
- to present the research findings intelligibly to all members of the community; and
- to motivate the different stakeholders from the five hamlets to proceed from conflicting perceptions to a shared view on the causes and consequences of their derelict irrigation system.

Would the people of CR-Palli avail themselves of the opportunity to utilize their local knowledge embedded in the above conflicting assumptions (Figure 10.3) and explore different ideas for future development? And finally, could shared perceptions, maturing in an interaction between indigenous and external knowledge, contribute to the empowerment of the community?

Establishing an arena for interaction

First, we set out to discuss with villagers the eco-history of the tank. Towards this goal, the research team distributed a written account of the history and the use of tank irrigation by the village. These accounts were presented orally in each of the five hamlets to introduce the research feedback process.

Second, the team presented an artist's poster-size impression of a typical South India watershed to probe people's own understanding of surface water irrigation. This was accompanied by a set of photographs of the important elements of the CR-Palli irrigation system. We invited the audience to relate these photographs to the watershed poster and in so doing to explain their ideas about the system and watershed. We refer here to John Campbell (2002) who draws our attention to the risks involved in applying scientific data on space, gained with GIS, without relating them to underlying conceptions of the actual users of the space.

Finally, we asked people to discuss the process of resource degradation on the basis of the common understanding established in the above exchange. Towards this end, the feedback team displayed a large diagram (corresponding to Figure 10.3) summarizing all the main explanations for the deterioration of the tank. We invited the gatherings to select from their own ranks a local expert panel of men and women to discuss 'water', and approached this panel for clarification of issues and resolution of conflicting interpretations. Research feedback thus helps to progress from agreeing on 'what is' to a shared understanding of 'why is it'.

Responses to the research team's feedback and audience participation were very positive, evidencing a deeply felt need for a better understanding of changes in the environment.

Reconciling diverse rationalities and multiple stakeholders

The village workshops agreed with the role of the watershed as catchment and the relationship between percolation of surface water and groundwater availability (Kumar 1994). They revealed detailed knowledge on the assumed course of groundwater streams. Discussion of causal factors showed that deterioration was thought to result from multiple factors.

In discussions, especially when visiting the tanks, a core rationality emerged governing attitudes toward water that is embedded in deep-rooted belief systems that consider water a resource controlled by non-human forces (Karanth 1995). The community enters into contact with these forces by making offerings at a small shrine on the tank bund. They appease these forces, for example, by discharging broken grinding stones over the northern boundary of the village, and in extreme cases, challenge them to bring rain by carrying out a mock burial in the forest. Can we really understand local water management practices without considering these rites and the related worldviews?

The feedback sessions also showed the extent to which multiple stakeholders' interests informed different definitions of the key problems and the proposed solutions. The team promoted the exploration of the various stakeholder perceptions through the following methods:

- Inviting participants to discuss the potential of an externally funded desilting operation[6] to solve the problem of the sandy silt in the tank.
- While acknowledging that erosion of the barren catchment had filled the tank with sand at an accelerated rate, focussing discussion on the deforestation of the hill slopes.
- Relating the reasons behind the discontinuation of traditional community desilting to changing village leadership, and comparing past and present village governance of local natural resources.

Any statement is closely linked to the political perspective of the stakeholder. Should researchers challenge such statements with their own views of unsustainable resource management? The research team sometimes faced a delicate task: direct contradiction could prove unproductive, or worse still, it may result in people losing face in public. We elevated the exchange of views to the generic level by asking such questions as: 'If the village had continued to protect the catchment and de-silt the tank regularly, how would the watershed look today?' Such questions opened the door to exploration of 'what actually happened' in the feedback rounds, putting the issue of village governance in the focus.

'Good village governance' in the past

Management of common resources in Indian villages traditionally has been an aspect of village governance (Wade 1988; Srinivas 1976). Before the abolition of the hereditary office of village headman in the 1960s, the community of CR-Palli was also under the leadership of a member of the dominant landholding caste, the Kapu-Reddy. Elders of the prominent lineage of the same caste constituted the panchayat, the village council. The old leader recalled to us his functions and responsibilities regarding the tank irrigation system. He regulated access to and use of the forests to optimize water harvesting in the upper catchment and organized people to fight forest fires during the dry season. He mobilized village labour to maintain the bund and the spill over canals and coordinated regular desilting of the tank to maintain its storage capacity. Finally, water distribution in the command area was under the supervision of two water overseers (neeruganti), both accountable to the headman. It was the headman's duty to settle any conflicts over use and maintenance of the tank at all levels.

The headman also played a crucial role in all major rituals linked to the agricultural production cycle. The Gangamma Puja is one example (Karanth & Ramaswamy 1998). This festival usually takes place when water starts flowing into the tank from the feeder channels. Village women, lead by women from the headman's house, proceed to worship the deity presiding at the tank bund. The officiating priest sacrifices an animal to appease her. The event only becomes ritually meaningful with the participation of the headman and his family. Thus in village rituals the traditional headman reaffirmed the collective responsibility of the village for its natural resources. Yet, the ascribed status also offered the dominating caste privileged access to natural resources and potential abuse of them.

Village governance in transition

The abolition in 1963 of the hereditary village offices, including that of the headman, was a first step towards more democratic panchayat councils. The process culminated in the recent 73rd Constitutional Amendment, which seeks to empower the gram sabha and create new opportunities for people to participate in rural development. The traditional jajmani obligations that linked landholding and service castes ritually and economically are also disappearing rapidly. The changes in leadership are considerable (Table 10.1). In CR-Palli, the elections in 1996 brought for the first time a member of the service washerman (Chakali) caste into power.

Table 10.1 Village leadership in transition

'Traditional Village Leader'		'Modern Village Leader'
Hereditary function of dominant landholding caste	**Power base**	Constitution with caste reservation and party affiliation
Autocratic and tradition based	**Style of functioning**	Bureaucratic and context-oriented
Unchallenged leadership with sanctioning power	**Role in internal conflict solution**	Formal authority related to public property, limited sanctioning power
Mobilizing local resources: labour, cash and kind	**Role in maintaining common pool resources**	Tapping external resources: Contracts, subsidies, grants
Central	**Role in village rituals and festivals**	Marginal
Long term, within generation cycles	**Time horizon**	Short term, 5 years election cycle

Village leadership in India today is a context specific blend of traditional and new forms of village governance (Ramaswamy et al. 2000). Behind the scenes the influence of traditional leaders is still evident. Depending on the issues, decision making at the village level may be the responsibility of the elected leader or the de facto leaders who command power by virtue of superior economic resources, connections to members of parliament, state legislatures, and government officials (Roy Chaudhury 1998). A significant impact of the emerging leadership pattern regarding natural resource management, is that:

> In the past, effective leadership was perceived as the capacity to mobilize the community and to pool indigenous resources. In the emerging pattern of leadership the sustainable management of natural resources has lost priority in favour of the efficient tapping of external resources from government and non-government sources (Ramaswamy 2000: 2).[7]

Such development goes hand-in-hand with an increasing orientation of rural livelihoods towards the outside world, beyond the village.

Common pool resources: from common property to open access

Farmers have increasingly turned from tank to pump irrigation. Technically, this is a move from surface water management to groundwater extraction (Olson 1965; Ostrom 1990). In terms of resource management, however, it involves a move from a common property to an open access regime (Agarwal & Narain 1997). During the feedback session, farmers were invited to recall the history of water management in their village. We supported this session with visuals showing the impact on groundwater streams. The impact of unregulated access to water was highlighted by the changing use of a prominent well in the village, situated in the tank bed. Formerly operated by a bullock driven matti system,[8] today the well is crowned with not less than a dozen diesel pumps owned by households of the landowning Kapu-Reddy caste in an attempt to satisfy their increased water requirements.

The trend is to sink deep tubewells in private fields, operated by electrical pump sets. The proportion of India's total irrigated area of approximately 50 million hectares, irrigated by wells and tubewells, increased from 38 per cent to 53 per cent between 1970 and 1993.[9] Figure 10.4 illustrates the dramatic increase in diesel and electrical pump sets over the last half century in Karnataka,[10] on the Deccan Plateau.

The causes of the expansion of wells are several. There are subsidies available for sinking tubewells and installing irrigation pumps, and there have been several programmes of large-scale deepening of wells in times of droughts. There is also cheap or even free access to electrical energy encouraging the shift from traditional food crops to irrigated cash crops (Kerr et al. 1996).

The move from common property management of water to an open-access regime based on groundwater use 'relieves' communities from holistically caring for catchments and integrated water management. The change to private extraction also reflects the disintegration of traditional village institutions (Wade 1988). A key biophysical concern is that the groundwater option is leading people to overlook the fact that water is a finite resource (Mönch 1998). Yet groundwater bodies regenerate each monsoon according to the percolation capacities of rock strata and also to human efforts in harvesting of surface water. With this background, the prospects of returning to sustainable community water management seem dim.

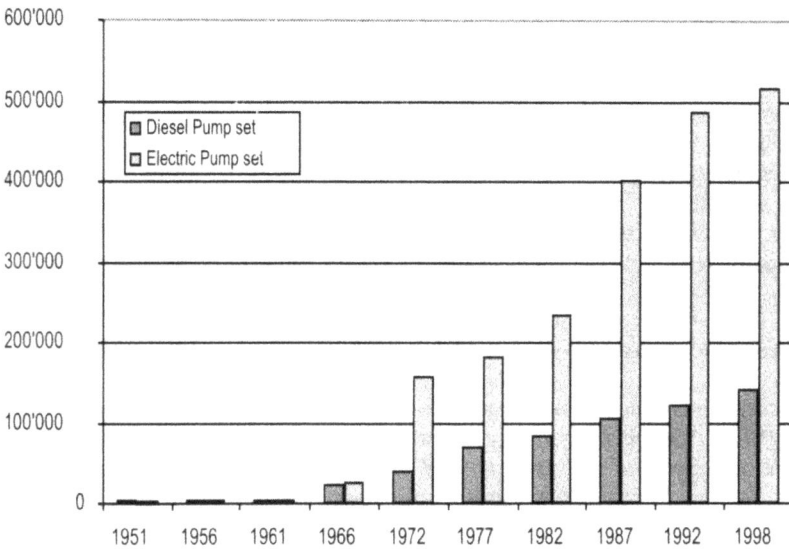

Source: Directorate of Economics and Statistics, Bangalore, Karnataka.

Figure 10.4   Diesel and electric pump sets installed in Karnataka, India

From research feedback to 'follow up'

Throughout the feedback sessions water remained a key issue for the village representatives. At the conclusion of the feedback, all five hamlets identified options for change in the field of natural resource management. After a long debate in search of clarity regarding the different roles of researchers and development agents the research team committed itself to:

- organize a fact finding trip for a village delegation to a watershed rehabilitation project where people's participation in planning and implementation was successful;
- investigate the state of the regional irrigation project which would have linked the CR-Palli tank with a perennial water source; and
- contract an expert in community forestry to advise CR-Palli on the legal options in rehabilitating the upper watershed catchment.

The last point in particular relates to the dissolution of institutions observed in many Indian villages. Enquiries showed that the five hamlets of CR-Palli would not be prepared to collaborate in reforestation, because of relations characterized by mutual distrust. This indicates the problems the 73rd Constitutional Amendment faces in instituting responsible panchayat village governments capable, for example, of watershed development (Tripathi 1997). Such trends to institutional disintegration and individual livelihood orientations within communities highlight the potential role of NGOs in rebuilding a stock of 'social capital', understood as the capacity to combine efforts to achieve a common goal.

Sustainable resource management at the village level requires a critical amount of collective action and responsibility. Social capital is a resource built in learning interactions (Coleman 1990; Falk & Kilpatrick 1999), to which our workshops aimed to contribute. The CR-Palli hamlets agreed that dividing the catchment into hamlet associated zones of responsibility or leasing out tree plots to individual households remain possible options in the present climate of mutual distrust.

Indigenous knowledge systems and forms of organizational learning

The indigenous knowledge identified with rural livelihoods and natural resource management has been to a large extent generated by experiential learning over generations, enriched and amalgamated over time with external knowledge (Kolb 1984). We can distinguish two sources of knowledge: that learnt within the context of the indigenous knowledge system (Thrupp 1989) and that learnt through interaction with external sources of knowledge. The sustainable use of water also features an institutional set-up involving organizational learning.

We have sketched therefore a model of a local knowledge portfolio that draws upon the work of Nonaka and Reinmöller (1998). In an attempt to understand the cultural dimensions of the endogenous generation of knowledge in Asian Economic Development, they advanced a model with four modes of knowledge creation.[11] In doing so, they draw upon the familiar distinction between tacit and explicit knowledge. Tacit knowledge is experience-bound, either individually internalized or collectively shared. Explicit knowledge is exchanged through dialogue, making it accessible for systematized and shared understanding. A society can promote effective learning by paying attention to the links between spheres of tacit and explicit knowledge by means of a 'learning spiral' (Nonaka 1994). Human decision making simultaneously features both explicit and tacit knowledge, contributing to what may seem 'irrational' to outside observers. It often appears beyond the capability of the possessor to make implicitly learned knowledge explicit, i.e., to communicate it (Reber 1993).

a) Assumption on structure of a knowledge portfolio and its areas of knowledge conversion

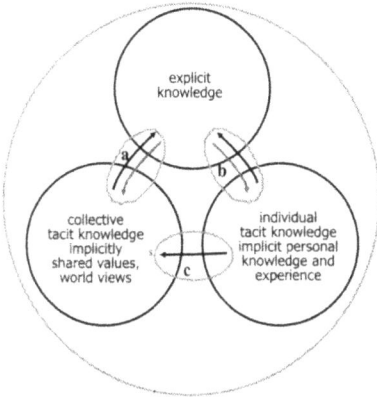

Endogenic learning through knowledge
conversion between spheres of knowledge

a: between explicit and tacit collective knowledge
b: between explicit and tacit individual knowledge
c: between the spheres of tacit knowledge

b) Learning through interactions with external knowledge systems

e.g. community   e.g. NGO

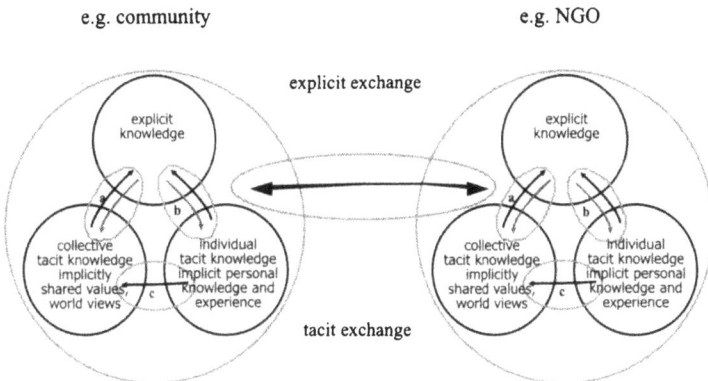

Figure 10.5 Knowledge portfolio

The model in Figure 10.5 proposes three modes of endogenous knowledge conversion. The conversion of tacit knowledge into explicit knowledge occurs, within limits, for example, when consciously addressing implicit values, traditions and worldviews, such as those governing the use of a natural resource such as water or forest. Worldviews are, to a considerable extent, embedded in tacit knowledge. For example, investigations into the perceptions of 'forest' by Bhil farmers in the Panchmahal District of Gujarat (Pathak and Ahmed 1995) revealed that the physical boundary between agricultural land and 'forest' also reflects a psychological separation between a familiar inner world, the cultivated land, and the jungle, perceived as vaira, the uncultivated outer world – the abode of tigers and evil spirits.

Reafforestation is hence not just a matter of introducing external knowledge, but of also addressing indigenous sources of tacit knowledge and psychological well being.

The practical application of new explicit knowledge again generates new individual tacit knowledge. This takes place, for example, when farmers start cultivating a new crop, transferred to them by extension agents as explicit knowledge. Its integration into their cropping system extends their body of personal and individual experience that also draws from tacit knowledge already in existence. Personal know-how may become 'embodied knowledge'.

Tacit knowledge shared within a community differs from collective explicit knowledge, codified rules and regulations (Baumard 1999). Collective tacit knowledge is a continuously emerging product of socialization into a common livelihood. It relates to 'social knowledge' as its sharing contributes to a sense of belonging. Village and caste based rituals performed in the course of the yearly cropping cycle powerfully express and reconfirm shared tacit knowledge. An underlying dilemma becomes obvious: while we may welcome the erosion of a traditional platform that supports undesirable dominance of upper castes, we witness at the same time how valuable social capital, understood as the ability to unite for collective action, declines.

Ignoring indigenous communication patterns and channels can result in ineffective development programmes (Mundy & Compton 1995). We need beware of the model confusing our perception of a society's knowledge portfolio. A caste based village society in India obviously structures its knowledge portfolio differently from a modern urban society under strong Western influence. The model also invites us to recognize the extent to which external factors influence indigenous knowledge generation and communication.

The three knowledge spheres, separated graphically in the knowledge portfolio, do not exist independently of each other but are interlinked when seen in the local context. It is these links between the tacit and explicit, between the spoken and the unspoken, that make it dangerous to treat indigenous knowledge as 'fungible' goods, freely convertible between different cultural contexts.

Learning for development within an indigenous knowledge system

The provision of research feedback has potential to promote endogenous learning within the boundaries of existing indigenous knowledge and to identify 'the tacit sources of knowledge' (Baumard 1999: 54). Our research feedback to the communities in Andra Pradesh and in Karnataka acted to move tacit knowledge into the realm of explicit knowledge, thus generating a learning process within the existing local knowledge portfolio. The feedback dealt not only with people's hypotheses regarding tank irrigation but also their attitudes towards catchment re-afforestation. It tackled underlying attitudes to migration, and motivated people to reflect on past and present forms of village governance. It encouraged people to share their painful experiences with the governmental watershed project. It made tacit knowledge of various fields accessible in positive dialogue towards shared understanding and action. The open acknowledgement of loss of faith between the hamlets in collaborating to re-stock the degraded upper catchment made this tacitly shared knowledge explicit paving the way to explore other options for catchment rehabilitation.

Assisting a community to make tacit individual and collective indigenous knowledge accessible for communication may initiate, or at least further dialogue on development issues. By promoting commonly shared perceptions on natural resource development, such as forests, irrigation tanks and pastures, a community gains in bargaining power for interactions with external development agents – be they government departments or NGOs. It generates awareness of best practices in land use, water management etc., in addition to mobilizing a community's existing stock of social capital. Research feedback creates a sense that one's voice matters and hence may promote local empowerment. As Sillitoe (2002a) puts it: 'Any involvement in research to inform development implies intervention'. In providing feedback to local communities researchers thus consciously assume the unavoidable role of knowledge brokers.

Learning through interactions between knowledge systems

Research feedback also creates a platform for interaction between two different knowledge systems. If we accept that learning is best facilitated in an environment where there is a dialectic tension and conflict between the immediate and concrete experience and analytical detachment (Kolb 1984), we can safely assume that the provision of research feedback has the potential to generate a productive learning environment of the above type. In the feedback situation illustrated in this paper, the farm community would be seen as associated mainly with 'immediate and concrete experience', and the research team with 'analytical detachment'.

Communication between stakeholders should be as equal partners, to create a 'dialectic tension' favourable to learning. Yet farmers and researchers hardly meet on egalitarian terms. This is a major stumbling block to effective learning. Furthermore, there are large inequalities within villages themselves that also hamper collective learning.

Expert knowledge carries status, prestige and power. The distortions experienced in interactions with expert knowledge systems indeed reflect the relative power balance between different knowledge traditions (Sillitoe, 2002a: Figure 7). It is thus essential, but maybe not sufficient, to acknowledge and grant status to indigenous knowledge before challenging indigenous experience with expert knowledge. Participatory Rural Appraisal (Chambers 1994; Schönhuth & Kievelitz 1994) has initiated a breakthrough in this regard, which we need to follow through, and also with research strategies responding to the demands of development collaboration. During the feedback sessions of the RLS project we:

- displayed indigenous knowledge accounts prominently on posters in the village 'open space' (Owen 1992) acknowledging local ownership of it and promoting the esteeming of existing stocks of knowledge;

- extended respect and gave roles to indigenous knowledge bearers – both women and men – in the sharing of research findings, for instance, by forming and addressing local expert panels;
- referred to carefully drafted life stories of local residents to convey changes in livelihood systems and relate these to natural resources;
- created opportunities for interaction between different communities by arranging for farmers from CR-Palli to visit a participatory watershed development project elsewhere.

In such an atmosphere, researchers may productively – but not degradingly – challenge local perceptions by first addressing them within the framework of prevailing indigenous knowledge, as was done in the case of conflicting assumptions among various stakeholders regarding watershed deterioration.

Providing research feedback has the potential to further knowledge integration for planned social change (see Purcell & Akinyi Onjoro, 2002). Certainly, meaningful knowledge integration does call for long duration research and context related methodologies, as these authors rightly state. Yet, the potential of knowledge integration through participatory research might be limited wherever people's path to learning is predominantly an experiential one. In other words, the contribution of anthropologists to knowledge integration might be more effective where research is linked to action in development implementation.

Feedback between 'research' and 'social activism'

There was heated debate within the RLS research team about engaging in research feedback, with some strongly against embarking on such an 'adventure' and others in favour of a new challenge to development research. The former asked: 'Does a researcher risk drifting into social activism when providing research feedback?' Indeed, we all agreed that development needs activists, – persons prepared to go beyond mere observing to actually fighting social injustice. There is no dearth of impressive examples of influential social activists in India, such as the Narmada dam conflict (Baviskar 1995), the dalit movement (Charsley & Karanth 1998), and various farmer movements (Nadkarni 1987). But what is the role of academics engaged in action oriented, participatory development research?

Stakeholder participation pushes the researcher to the threshold between analysis and action. During feedback, when researchers engage in critical discourse of development options emerging within a community, there is a danger of crossing this threshold. Effective 'participatory development research' demands that we face up to it and not retreat into conventional, personally detached participant observation (Wright & Nelson 1995). Must researchers therefore become activists to contribute effectively to development, moving from words to deeds? Or, is role clarity – and hence role separation – more effective? Villagers in CR-Palli expected empathy for their request to de-silt the tank rather than an academically detached analysis of its futility without rehabilitating the catchment. Should researchers still refrain from social activism even when people make knowledge of mismanagement explicit, as with watershed committees?

A researcher may be led into activism by good intentions. The danger is that these may fade with time and when funding ceases. The researchers then leave people in villages behind who have put their faith in future support by outsiders, and may even have made themselves vulnerable to local power brokers. When activism and research become confused, research findings can easily be labelled by decision makers as biased and unreliable, nurturing the suspicion that the data may not have been transparently gathered but purposely tailored to support a chosen party in a conflict. Research risks losing its credibility and scope to inform the development process. Research must be free to both support and challenge activist movements in as unbiased a way as possible. But how can development researchers avoid returning to 'quarry type' field research, mining data from rural communities for academic reports?

One alternative is to 'hand over the baton' to partners – NGO or government agency – engaged in development project implementation. This depends on a development partner existing in the research region who is prepared to assume the delicate role of interfacing between researchers and community. Such a partner must be able to further development based on insights gained through research – in other words, be capable to contribute to the research agenda and to integrate research findings into implementation.

Another alternative is to combine research with action, in a participatory sequence of joint reflection-action-reflection. Meaningful action research requires researcher involvement in the conception and moderation phases as well as during analysis and evaluation, (e.g., participatory technology development where innovation is the product of an interaction between indigenous and external knowledge systems (Haverkort et al. 1991; Röling 1992)). On-farm seed developments are examples of such successful interactions where research feedback is built into the research process (see Rhoades & Bebbington 1995).

The risk of confusion between participatory research and social activism becomes evident during the provision of research feedback to a community. The risks should be taken as a challenge rather than an excuse not to share the results of research, which are dependent on a community's cooperation. Participatory research also potentially strengthens cultural identity and contributes to a community's ability to cope with social change (Robinson 1996). The provision of effective and responsible feedback requires considerable investment in time and funds, and should be integrated into the drafting of research budgets and schedules. Research feedback should be a joint responsibility to the benefit of all partners. And finally, actor-oriented research can best support rural development if part of a network of communities and development agents. Such arrangements require clarity of roles, mutual accountability and shared ownership of research.

Notes

1   This research was sponsored by the Swiss National Science Foundation under its Priority Programme, 'Environment: Module 7'. Research partners were drawn from different development and research Institutions in India. This paper reports the results of collaboration between the authors from the Institute for Social and Economic Change,

228 Investigating Local Knowledge

Bangalore, and of NADEL-ETH (Federal Institute of Technology), Zurich. We gratefully acknowledge the comments offered by M.N. Srinivas at an earlier presentation of the paper.

2   After independence in 1947 the State Governments of India transferred responsibility for maintaining larger village tanks to various government departments.

3   The *jajmani* system regulates the relationship and mutual obligations between landowning castes and service castes in an Indian village context.

4   The authors wish to express their gratitude to Dr. K. G. Gayathri Devi, Ms. Rupashri and Ms. Meera Murthy for their invaluable role in this respect.

5   CR-Palli encompasses a cluster of five hamlets and comprises 270 households belonging to 15 different castes or *jatis* (Dumont 1980). Social stratification is reflected in the settlement pattern and type of housing (Ramaswamy et al. 1996).

6   According to an action plan of the Minor Irrigation Department, 40,800 tanks in the neighbouring state of Karnataka, with a storage surface of 900,000 hectares have been identified for revival (Kumar 1999). The plan is based on successful desilting of tanks on an experimental basis in Kolar district, one of the research areas of RLS Project. According to Kumar, the silt provided for their fields mainly generates the farmer's motivation, which makes sustainable use of tanks doubtful.

7   During an initial and exploratory phase the RLS Research developed a number of guiding assumptions for a more focused research process in a second phase. Transitions of village leadership became such a research focus. Research feedback was an element of both phases, first in CR-Palli in Andra Pradesh, then in the village of Kodipalli in Karnataka (Ramaswamy et al. 2000: 3).

8   *Matti* irrigation is a bullock powered water lifting system operating with a leather bucket. The area irrigated by one 0-seldom exceeded three acres.

9   Statistical Outline of India 1997-8, TATA Services Limited, Department of Economics and Statistics, Bombay House, Mumbai.

10 Data drawn from various sources: (a) A Brochure on Irrigation Statistics in Karnataka, 1980-81 to 1993-94, Directorate of Economics and Statistics, 1995, Bangalore. Pp. 62-75; (b) Statistical Abstract of India 1997, Directorate of Economics and Statistics, New Delhi; and (c) Government of Karnataka, Livestock Census, 1997, Karnataka State, Directorate of Animal Husbandry and Veterinary Services, Bangalore.

11 The keywords for the four modes are: *Externalization/Articulation* (transforming tacit into explicit knowledge); *Internalization* (transforming explicit into tacit knowledge); $0-$ (formation of collective tacit knowledge); and *Combination/Systematization* (integrating different forms of explicit knowledge) (Nonaka and Reinmöller 1998).

References

Aurora, G. S. 1999. *Two villages in semi-arid tropics of India: A comparative study of Mahudi and CR-Palli.* Zurich: NADEL; Anand: IRMA; and Bangalore: ISEC. Mimeo.

Agarwal, A. and S. Narain. 1997. *Dying wisdom: rise, fall and potential of India's traditional water harvesting systems.* New Delhi: Centre for Science and Environment.

Baumard, P. 1999. *Tacit knowledge in organizations.* London: Sage Publications. (Originally published as Organisations Déconcertées: La gestion stratégique de la connaissance. Paris: Masson, 1996).

Baviskar, A. 1995. *In the belly of the river: Tribal conflicts over development in the Narmada valley.* New Delhi: Oxford University Press.

Carney, D. (ed.) 1998. *Sustainable rural livelihoods: What contributions can we make?* Papers presented at the Department for International Development's Natural Resource Advisors Conference, July 1998, DFID, Department for International Development.

Chambers, R. (ed.) 1979. Rural development: Whose knowledge counts? *IDS Bulletin.* 10(2).

Chambers, R. 1994. The origins and practice of participatory rural appraisal. *World Development.* 22(7): 953-969.

Chambers, R. 1995. Poverty and livelihoods: Whose reality counts? *IDS Discussion Paper* 347.

Charsley, S.R. and G.K. Karanth. (eds.) 1998. *Challenging untouchability: Dalit initiative and experience from Karnataka.* New Delhi: Sage Publications.

Coleman, J., 1990. *Foundations of social theory.* Cambridge, MA: Belknap Press of Harvard University Press.

Cornwall, A. and R. Jewkes. 1995. What is participatory research? *Social science and medicine* 41(12): 1667-1676.

Dixon, J.A. 1997. Analysis and management of watersheds. In *The environment and emerging development issues.* Vol. 2. (eds.) P. Dasgupta and K.G. Mälur. Oxford: Clarendon Press.

Dumont, L. 1980. *Homo hierarchicus: The caste system and its implications.* Chicago: University of Chicago Press.

Falk, I. and S. Kilpatrick. 1999. *What is social capital? A Study of interaction in a rural community.* Paper 5/1999 in the CRLRA (Centre for Research and Learning in Regional Australia) Discussion Paper Series.

Haverkort, B., V.D.J. Kamp and A. Waters-Bayer. 1991. *Joining farmers' experiments: experience in participatory technology development.* London: Intermediate Technology Publications.

Hinchcliffe, F., J. Thompson, J. Pretty, I. Guijit and P. Shah. (eds.) 1999. *Fertile ground: The impacts of participatory watershed management.* London: IT Publications.

Högger, R. 2000. *Understanding livelihood systems as complex wholes.* Zurich: NADEL. Mimeo.

Jodha, N.S. 1990. *Drought management: The farmers' strategies and their policy implications.* Issues paper no.21. London: IIED.

Jeffery, R. & N. Sundar. (eds.) 1999. *A new moral economy for India's Forests? Discourses of community and participation.* New Delhi: Sage.

Karanth, G.K. 1995. *Surviving droughts.* Bombay: Himalaya Publishing House.

Karanth, G.K. and V. Ramaswamy. 1998. The threshing floor disappears: Kodipalli's livelihood in transition. Bangalore: ISEC. Mimeo.

Kerr, J.M., N.K. Sanghi & G. Sriramappa. 1996. *Subsidies in watershed development projects in India: Distortions and opportunities.* Gatekeeper Series No.61, London: IIED.

Kolb, D.A. 1984. Experiential learning: Experience as the source of learning and development. New Jersey: Prentice-Hall.

Koppers, W. 1948. *Die Bhil in Zentralinden.* Horn-Wien: Verlag Ferdinand Berger.

Kuhn, B. 1998. *Participatory development in rural India.* New Delhi: Radiant Publishers.

Kumar, P.P.D. 1994. Farmers are Engineers: Indigenous Soil and Water Conservation Practices in a Participatory Watershed Development Programme, PIDOW. Bangalore: SDC. Mimeo.

Kumar, R.V. 1999. Tanks in the state to be revived. *The Hindu.* Bangalore, February 24.

Mundy, P.A. and J.L. Compton. 1995. Indigenous communication and indigenous knowledge. In *The cultural dimension of development: Indigenous knowledge systems.* (eds.) D.M. Warren, L.J. Slikkerveer & D. Brokensha. London: Intermediate Technology Publications. 112-123.

Nadkarni, M. V. 1987. *Farmers movements in India.* Bombay: Popular Prakashan.

Nelson, N. and S. Wright. (eds.) 1995. *Power and participatory development: Theory and practice.* London: Intermediate Technology Publications.

Nonaka, I. 1994. A dynamic theory of organizational knowledge creation. *Organization Science.* 5(1): 14-37.

Nonaka, I. and P. Reinmöller. 1998. *The legacy of learning: Toward endogenous knowledge.* Creation for Asian Economic Development. Jahrbuch: WBZ.

Olson, M. 1965. *The logic of collective action, public goods and the theory of groups.* Cambridge: Massachusetts: Harvard University Press.

Oommen, M. A. 1995. *Devolution of resources from the State to the Panchayati Raj institutions: Search for a normative approach.* Occasional Paper Series, No 18. New Delhi: Institute of Social Sciences.

Ostrom, E. 1990. *Governing the common: The evolution of institutions for collective action.* New York: Cambridge University Press.

Owen, H. 1992. Open space technology: A user's guide. Potomac: Abbot Publishing.

Pathak, A. & S. Ahmed. 1995. *Forests, the abode of evil: Positing Mahudi against ecological prudence.* Anand: Institute of Rural Management. Mimeo.

Pretty J. N. & I. Scoones. 1995. Institutionalizing adaptive planning and local level concerns: Looking to the future. In *Power and participatory development: Theory and practice.* (eds.) N. Nelson & S. Wright. London: Intermediate Technology Publications. 157-169.

Ramaswamy, V., G.S. Aurora & G.K. Karanth. 1996. *Rural livelihood systems and sustainable resource management, a report from an Andhra Pradesh village.* Bangalore: ISEC. Mimeo.

Ramaswamy, V., G.K. Karanth, R. Baumgartner & R. Högger. 2000. *Emerging rural leadership and sustainable management of natural resources: Evidence from two south Indian villages.* Bangalore: ISEC. Mimeo.

Reber, A.S. 1993. *Implicit learning and tacit knowledge: An essay on the cognitive unconscious.* Oxford Psychology, Series No 19. Oxford: Clarendon Press.

Rhoades, R.E. 1998. *Participatory watershed research and management: Where the shadow falls.* Gatekeeper Series, No. 81. London: IIED.

Rhoades, R. & A. Bebbington. 1995. Farmers who experiment: An untapped resource for agricultural research and development. In *The cultural dimension of development: Indigenous knowledge systems.* (eds.) D. M. Warren, L. J. Slikkerveer & D. Brokensha. London: Intermediate Technology Publications. 296-307.

Robinson, C. D. W. 1996. *Language use in rural development: An African perspective.* Berlin and New York: Mouton de Gruyter.

Röling, N. 1992. The emergence of knowledge systems thinking: A changing perception of relationships among innovation, knowledge process and configuration. *Knowledge and Policy: The International Journal of Knowledge Transfer and Utilization* 5(1): 12-64.

Roopi, S.G. 1996. Watershed in corruption: half of the project sum misappropriated. *Deccan Chronicle.* Hyderabad, December 2.

Roy Chaudhury, S. 1998. *Leadership, natural resources and rural livelihood in transition.* Bangalore: ISEC. Mimeo.

Schönhuth, M. & U. Kievelitz. 1994. *Participatory learning approaches. Rapid rural appraisal, participatory rural appraisal: an introductory guide.* TZ Verlag Rossdorf: Deutsche Gesellschaft für Technische Zusammenarbeit (GTZ).

Scott, R.A. & A.R. Shore. 1979. *Why Sociology does not apply: A study of the use of sociology in public policy.* New York: Elsevier.

Srinivas, M.N. 1976. *The remembered village.* New Delhi: Oxford University Press.

Thrupp, L.A. 1989. Legitimizing local knowledge: 'Scientized packages' or empowerment for Third World people? In *Indigenous knowledge systems: Implications for agriculture and international development.* (eds.) D.M.Warren, L.J. Slikkerveer and S.O. Titilola. Ames: Iowa State University Press. 138-153.

Tripathi, P.M. 1997. Panchayati Raj: Expectations and frustrations. *Wasteland New.* Aug-Oct: 14-18.

Vasavi, A.R. 1999. *Harbingers of rain: Land and life in South India.* Delhi: Oxford University Press.

Wade, R. 1988. *Village republics, economic conditions for collective action in South India.* South Asian Studies Series. Cambridge: Cambridge University Press.

Whyte, W.F. 1984. *Learning from the field: A guide from experience.* London: Sage Publications, Inc.

Wright, S. and N. Nelson. 1995. Participatory research and participant observation: Two incompatible approaches. In *Power and participatory development: Theory and practice.* (eds.) N. Nelson and S. Wright. London: Intermediate Technology Publications. 43-59.

Index

For Product Safety Concerns and Information please contact our EU
representative GPSR@taylorandfrancis.com
Taylor & Francis Verlag GmbH, Kaufingerstraße 24, 80331 München, Germany

* 9 7 8 1 1 3 8 3 5 6 1 8 4 *